Genetic
Hope for the Future or Pandora's Box?

—A new form of bacteria that can devour oil spills at sea, then disappear when that work is accomplished.

—A strain of cancer that would reach epidemic proportions around the world without the source of contamination being suspected.

—A way for scientists to understand and utilize genetic immunities to disease.

—Cloning of living organisms, including human beings.

—The scientific breakthroughs that led to the conception and birth of the first test-tube baby.

All of the above possibilities are among the multitude of factors that are now being weighed as the scientific community finds itself taking on a monumental responsibility along with its power to alter life.

BIO-REVOLUTION
DNA and the Ethics
of Man-Made Life

RICHARD HUTTON, a graduate of the University of California at Berkeley, is a free-lance writer with special interest in science and medicine. He has co-authored several medical textbooks, and has written and translated numerous articles on scientific subjects, as well as publishing science fiction stories.

Other MENTOR and SIGNET Books of Interest

BIO-REVOLUTION

DNA and the Ethics of Man-Made Life

by Richard Hutton

A MENTOR BOOK

NEW AMERICAN LIBRARY

TIMES MIRROR

New York and Scarborough, Ontario
The New English Library Limited, London

To Martha Sadler Dungey,
blessed with 46 of the best.

Library of Congress Catalog Card Number: 78-070067

 MENTOR TRADEMARK REG. U.S. PAT. OFF. AND FOREIGN COUNTRIES
REGISTERED TRADEMARK—MARCA REGISTRADA
HECHO EN CHICAGO, U.S.A.

SIGNET, SIGNET CLASSICS, MENTOR, PLUME and
MERIDIAN BOOKS are published *in the United States* by
The New American Library, Inc., 1301 Avenue of the Amer-
icas, New York, New York 10019, *In Canada* by The New
American Library of Canada Limited, 81 Mack Avenue, Scar-
borough, Ontario M1L 1M8, *in the United Kingdom* by The
New English Library Limited, Barnard's Inn, Holborn, London,
E.C. 1, England

First Mentor Printing, December, 1978

1 2 3 4 5 6 7 8 9

PRINTED IN THE UNITED STATES OF AMERICA

The recombinant DNA technology promises to give man power over nature of a more fundamental kind than that afforded by any other technology. . . . Hitherto evolution has always seemed as inexorable and irrevocable a process as time or entropy. Now at last man has a handle on the force that shaped him.

—*Nicholas Wade*

This world is given to us on loan. . . . My generation, or perhaps the one preceding mine, has been the first to engage under the leadership of the exact sciences, in destructive colonial warfare against nature. The future will curse us for it.

—*Erwin Chargaff*

Contents

Acknowledgments

I owe a measure of thanks to more people than I can realistically name in a few short paragraphs. I cannot say that this book would not have been finished without them; but without them, I could not have written this particular book. Among those to whom I owe the most:

Noel Gunther, who read through the entire manuscript, and whose blunt, incisive comments smoothed and tightened some of my more turgid, convoluted prose and logic.

Zsolt Harsanyi, who weaned me on the science of genetics, told me things I did not know, and explained to me what I would not have otherwise understood.

Hugh Thomas, whose sense of line and proportion are expressed in the illustrations accompanying the book.

Norton Zinder, Erwin Chargaff, and Daniel Callahan, who kindly consented to sometimes long, often exciting interviews.

Charles Weiner, Rae Goodell, and Mary Terrall of the Massachusetts Institute of Technology's Department of Oral History, whose fantastic dedication to collecting every scrap of published information and fanatically detailed interviews with scores of the participants in the DNA debate simplified my research immensely.

Maxine Losseff, whose efforts in locating research materials were a great help.

June Goodfield, who gave me selfless encouragement at some critical times.

Theda Muir, who typed and proofread large chunks of the final manuscript.

Ron Kutak and Stanley Hoppenfeld, whose friendship and support made the harder times easy.

Barry Lippman, my editor, for his capacity to combine genuine enthusiasm and deadly accurate criticism, and for his care in shepherding this manuscript through its publication.

Introduction

It may be the new Pandora's Box; it may let loose into the world a torrent of nightmares beyond our ability to control. Or it may clear a path to knowledge and technology so important that it will change the face of mankind. This is the controversy surrounding the most recent experimental manipulations of DNA, the building block upon which all life is based.

Scientists, researchers, militant consumers, environmentalists, concerned citizens, and the merely curious have staked out positions on both sides of the issue. Some can see the potential benefits of the research, and view the possibility of danger as remote, if not completely hypothetical. Others feel that the dangers, however distant, are so potentially catastrophic that they justify postponing the progress of the research or halting it altogether.

There is no question that the genetic research now being conducted has implications that, until recently, seemed safely tucked away in the upper reaches of science fiction. But the present debate is over real issues, and its outcome, beyond deciding how we will handle our new knowledge, will send repercussions throughout the scientific community, influencing the future of issues as broad as the right of free inquiry and the morality of research in general.

A cartoon published in *The New Yorker* not too long ago depicts a scientific laboratory at the moment of triumph at the end of an experiment. A squad of ravenous giant rats, emerging from experimental slumber, are ripping through the wire of their cages and escaping into the world. Researchers are fleeing madly in all directions. As they run for their lives, they are shouting, "Eureka! Eureka!"

This cartoon mirrors some of our preconceptions about scientists: they seem introspective, involved only in their work, unable to look up long enough to form a social conscience. The stereotype is simplistic. Yet it has been drawn often enough to enter our consciousness and to affect the way we view the relationship between science and society.

Gradually, over the past century, the pace of scientific research has accelerated. The bond between basic research and its applications, between the discovery of new information and the creation of technology that uses it, has also grown. As a result, research that once seemed unrelated to our lives has gained a direct and powerful influence in society. It is this close relationship between society's future and the implications of research that make it as important for us to examine scientific progress as it is for the researcher to recognize the needs, limitations, and requirements of society. And it is crucial that we remain objective.

Whether we wish to acknowledge it or not, science is a two-edged sword. One edge provides mankind with the means for improvement and advancement. By using it, we have been able to shed the ignorance of the Dark Ages and construct the scientific method, an orderly way of investigating and disseminating new information. We have experienced the Industrial Revolution, which has freed us from otherwise necessary drudgery and has given us time for leisure and the arts. In the nineteenth century, we began to gain a measure of control over disease and death, for as the rate of infant mortality dropped, life expectancy soared. And we have learned to delve into microcosms, into the psyche and the atom, initiating an organized analysis of the former and harnessing the latter for its potential energy.

But none of these benefits has come to us without cost. In each instance, a lack of foresight or a desire to rush headlong into applying our new knowledge has run us up against the second cutting edge of that scientific sword. And although we would not trade the scientific and technological advances of the last century and a half, we might well wish for second sight, for the ability to go back and do it all again more conscientiously, more carefully, with more of an eye toward the future. In addition to its benefits, what has this progress brought us? The Industrial Revolution, its processes unchecked and its consequences ignored, has led to an irreversible waste of natural resources. Their destruction has continued despite our awareness of the dangers. In medicine, death control without the necessary interpolation of comprehensive birth control has caused a population explosion which threatens to consume the earth's remaining natural resources and to strain to the limit its ability to provide food. Only now, when the problem has become overwhelming, has the need for careful population control entered upon our national consciousness; and because the consequences of overpopulation do not stop simply at political boundaries, it will do us little good at home if the world's problem continues unabated. Hand in hand with control of the atom has come nuclear weaponry and the danger it entails. The generation that invented the atomic, hydrogen, and plutonium bombs has surely left a legacy for the future that flirts with disaster. With brinkmanship a political reality, we have consigned future generations to live in constant awareness that the means for their destruction lies only as far away as the pressure of a human finger.

Now the science of genetics is faced with a biological revolution that presents us with the same dichotomy of risk and benefit. It is perhaps both ironic and fortunate that science has begun to uncover the secrets of DNA just as we realize the importance of considering what both this research and research in general mean. For the recent triumphs in genetics are not simple run-of-the-mill successes; both the research itself and its applications have potential as profound as any scientific discovery previously encountered.

In 1944, Oswald Avery first came to the conclusion that DNA was the storehouse of genetic information, that each long strand of DNA had, coded into it, all the inherited characteristics that make each individual form of life what it is. Avery's discovery focused, intensified, and accelerated the research that followed; today, just over thirty years later, researchers can slice a strand of DNA into pieces and recombine it with pieces of another type of DNA to form new hybrids. These experiments in recombinant DNA have permitted researchers to take a quantum leap forward in understanding the DNA storehouse, in mapping its codes, and in finding ways to alter the genetic construct of a living cell.

The research itself may sound remote and only indirectly connected with the way we live, but its potential for application is enormous. In medicine, for example, it is known that certain genes control single characteristics; there is a gene for hemophilia, another for sickle-cell anemia, a third that causes certain types of cleft palates. Soon scientists may be able to isolate the defective gene, remove it from the patient's gene pool, and replace it with a healthy gene, protecting both the patient and his offspring from deformity or disease. Other possibilities include creating antibiotics that attack specific diseases, breeding bacteria that produce insulin, growth hormones, and the clotting factor, and breeding viruses that can prevent or inhibit the growth of cancer.

Commercially, the potential is just as vast. In agriculture, it may soon be possible to alter the genetic structure of crops so that they can extract what nitrogen they need directly from the air instead of having to draw it out of the soil. Scientists are also attempting to breed bacteria that devour oil and will mop up after accidental spills at sea.

It sounds magnificent. But even though these concepts are nowhere near the stage of development where they can be put to practical use, they are nonetheless real possibilities.

Possible too, however, are the dangers. The most immediate one, and the one that touched off the original controversy, involves the basic research of recombining two different kinds of DNA. Risks in basic research are, in themselves, a new wrinkle; in the past, risks have grown out of applied research, from the technology developed as a consequence of new knowledge. In research with recombinant DNA, the basic experimentation itself hypothetically contains as much danger as its application.

Imagine, for a moment, a researcher working in his laboratory. He is well aware of the hazards of his work: the wholesale dividing of a strand of DNA, and the recombination of the pieces with DNA from another source, will create previously nonexistent hybrids. Most are harmless—perhaps all. But because of the danger of one of those hybrids escaping and freeing a new, untested, potentially dangerous form of life irreversibly into the biosphere, he is working in a sealed laboratory, one designed to minimize the chance of escape. Indeed, even the bacteria he is working with have been bred so that only one in one hundred million can live outside of the experimental environment for as long as it would take to transfer its genetic information to another, more viable host.

The scientist now begins an experiment he has successfully completed many times before. But this time, inexplicably, recombination takes place between the second source of DNA and a quiescent cancer gene in the host cell. The quiescent gene becomes active. A moment of carelessness, an accident, a mechanical breakdown in the laboratory's machinery, contrives to free the bacterium containing that gene. Before it dies, it gains access to the scientist's body through his mouth and exchanges its own carcinogenic information with a viable bacterium in his intestine. The bacterium replicates; the dangerous genetic information, now active, is passed from the scientist to his family and beyond. By the time the cancer is detectable, and by the time its origin is traced to the laboratory, an epidemic of cancer has begun, with no measures of containment either known or available.

While the chances of this scenario occurring seem remote to us now, not even those who advocate unrestricted research can assure us that it will not happen. In fact, during the last thirty years, there have been over 5,000 documented cases of laboratory-acquired infections in all types of research, with one-third occurring in laboratories equipped with special containment facilities. And at Fort Detrick, Maryland, where the government used perhaps the best facility for physical containment in existence to conduct experiments for biological warfare, there have been 423 accidents—escapes by the tiny biological inmates—leading to three deaths in the past twenty-five years. With a safety record like this in our best facility, we must certainly be concerned with the conduct of research in all the others. Work with recombinant DNA is expanding yearly. As hypothetical as our small scenario might seem, it is almost as inconceivable that an accident of some kind will not occur in one of the many facilities housing the experimentation that will spring up around the country in the near future.

The possibility of such serious risks arising from the conduct of basic research raises a series of questions, all of which must be answered before we can become thoroughly comfortable with recombination:

- The nature and extent of the risks must be clarified.

- The social implications of scientific accidents must be explored.

- Judgments must be made as to the acceptability of the risks—both in a societal and in a moral sense.

- Basic policy decisions have to be explored; it must be determined what kinds of conduct are appropriate for research, for the resultant technology, for the public, and for the decision-makers.

It is quite clear that the problems of recombinant DNA research do not simply stop with the basic research. Equally hazardous potential exists in its application. Despite current international restrictions, genetic research is almost tailor-made for experiments in biological warfare. Bacteria similar to those used in producing life-saving insulin might become havens for virulent forms of disease or, in a less lethal vein, for a gene carrying a code that could spark a national attack of dysentery.

Perhaps most frightening is the fact that our present state of knowledge makes it almost impossible to project where research now being undertaken will lead us. Along with the potential for dramatic, fundamental breakthroughs in almost every biological discipline goes the possibility that some discoveries may lead to an understanding of the exact processes that combine to form and create human life. And knowing what these processes are is only one step removed from experimenting with them, modifying them, and, ultimately, changing the very nature of man. At one bizarre extreme, it is not hard to imagine our gaining control of the mechanism that regulates intelligence or old age and manipulating it drastically enough to render the present form of our own species obsolete. Together, the benefits and risks add up to an obligation to talk about the future long before it arrives, and to talk about it in a way that involves not only the technical aspects of the research, but its moral implications as well.

Scientists and researchers are by no means unaware of the problems. Indeed, efforts to regulate recombinant DNA research actually originated from within the ranks of those performing the research. The debate went public from the beginning; the facts have been accessible, if disorganized and disputed, ever since. And the guidelines which have emerged from the first years of discussion, although preliminary, are far from cosmetic; drawn up by the National Institutes of Health (NIH), one of the major sources of funding for scientific research, they define precise parameters within which the more volatile forms of research may be conducted. Although, at present, the only penalty for not complying is the loss of funding, the guidelines have the considerable weight of moral suasion behind them. European countries involved in research into recombinant DNA have also generally accepted the NIH guidelines and are modeling their own along the same lines.

The international scientific community has further demonstrated its concern by taking part in a historic conference in Asilomar, California. The conference's purpose was to review the progress of recombinant DNA research and to discuss appropriate ways of dealing with its potential biohazards. The one hundred forty internationally recognized scientists who attended were able to subordinate the demands of their own research long enough to construct and approve the set of recommendations that the NIH used as the foundation for their guidelines. In doing so, they displayed a recognition of the realities of the bond between society and science and of their own responsibilities for preserving the relative freedom of inquiry that they enjoy.

Unfortunately, not all scientists show such wisdom. A lawyer who lectures frequently on the need for scientific responsibility tells a story which may or may not be apocryphal. He was once asked to address a group of scientists on the question of the ethics of recombinant DNA research. During his talk, he hit hard upon the issue of the scientist's fiscal responsibility for the consequences of his research, saying that the scientist, not society, carried the ethical burden of proof in cases of scientific risk. When he had finished speaking he asked for questions from his audience. One man in the back, obviously concerned about the financial implications of the lawyer's remarks, raised his hand. When he was recognized, he asked, "I work in a laboratory I built on my own property. Does that mean I am covered by my homeowner's policy?"

Few researchers have the insensitivity to remain this oblivious to societal pressures. But a strong and constant debate is raging over the degree to which a scientist, whose primary loyalty is to his research, should have to accede to the conflicting and often indecisive demands of society. To many, society's perspective is distorted by intuitive and emotional variables; and there is no way to reconcile those variables with the necessary impartiality that effective research requires.

This subliminal conflict between the divided loyalties of the scientist has taken the debate over recombinant DNA research from its natural confines and made it a test case for the more general relationship between science and society. The issues in both cases are similar, for whatever is decided about DNA will set precedent for analogous cases elsewhere. As we watch the controversy unfold, we are also watching, in microcosm, the course of coming controversies. The decisions to be made—as to how risks and benefits should be weighed, who will do the weighing, how guidelines will be enforced, and how scientists will react to their modified roles—are the same as those that must be made in the future.

With so much at stake there can be no unseemly rush to judgment. The researchers must continue to gather and evaluate technical evidence and disseminate their findings openly. In the meantime, we must devise a historical perspective which examines our past experience with technological innovation and scientific progress and weighs it against the lessons we have learned. Finally, we must determine how best to involve the public, the group that pays the price. Even though society should have a voice in deciding what and how much it is willing to risk, its influence should increase with the state of its knowledge, and not with the stridency of irresponsible advocacy by either side. Ultimately, mastering the facts is the key to a rational, realistic decision. The question of whether to support or restrain the new biological revolution is important enough to deserve that kind of attention.

ONE:

The Background

The Pioneers

It must have begun when someone, relaxing from a hard day of hunting or hauling water from a nearby stream, idly contemplated a herd of wild horses or a pack of wolves and realized that the horses produced other horses, and not cows, and that wolves generated little wolves, and not fish. Recognizing the existence of heredity was probably one of man's first scientific observations.

It must have become obvious soon after that this logic held true for all kinds of living things, including man himself, and that it was the act of mating between male and female that permitted us to reproduce. Making the connection between mating and heredity might well be one of the earliest examples of our ability to reason deductively.

Long before he began to record his observations and exploits in writing, man had learned primitive ways of manipulating genes. Heredity quickly became a tool, a way of improving life. It was one of the elements that made possible the shift from the nomadic, uncertain existence of the hunter to the life of stability and relative plenty of the farmer, where collection of food was a simple matter of planning, foresight, and a little luck. The rise of civilization—when man first settled in permanent dwellings—was due in large part to man's ability to control the inherited characteristics of animals and plants—to breed tame livestock and to modify wild plant life for cultivation.

These and other practical aspects of heredity—some of its hazards and potentials—became clear relatively quickly: The Old Testament is stocked with passages on the need for crop selection and the benefits of breeding pure strains of cattle. This newfound knowledge also permitted man to practice a limited kind of eugenics, almost instinctively protecting the human gene pool against deterioration; the nearly universal proscription against marriage between close blood relatives and the Greeks' quaint practice of ridding themselves of "defective" offspring by flinging them from cliffs were both crude ways of shielding mankind from some of the most obvious undesirable inherited characteristics.

The problem was that everybody could see that heredity existed, but nobody could understand how it worked. Because its secrets are all contained in that microscopic entity, the cell, and because people had no idea that cells even existed, much less that heredity was tied up in their function, the philosophers could only guess at what was going on. Some of the explanations were marvelous; some came close to the truth. Aristotle, with his theory that the male's semen contains the plans for the offspring, advanced to the edge of understanding, while others developed the poignant and logical concept of the homunculus, the tiny, wholly formed midget that lived in the sperm, waiting only to be inserted into the womb before growing. Despite the hundreds of hypotheses that surfaced, foundered, and were raised again, the understanding of heredity remained at a standstill for thousands of years. Techniques for its use were refined; people discovered that mating a horse with a donkey could produce a wonderful, though sterile, pack animal, the mule. But the next real step in the understanding of heredity was destined to wait for a shy, curious monk who puttered through a pea patch in Brno, Moravia.

In 1865, Gregor Mendel presented his work, "Experiments in Plant Hybridization," to the Brno Society of Natural Science. Working with pea plants, he had devised a set of principles which explained why offspring imitated their parents in appearance and which answered some of the larger questions of genetic inheritance. In his experiments, Mendel had traced the consequences of breeding different strains of peas—each with distinct characteristics, such as seed color—to each other. First, he found that each pure strain bred true, that two plants with yellow seeds would produce other yellow-seeded plants. But if plants with yellow seeds were crossed with those having green seeds, the offspring would acquire the characteristics of both parent plants in a thoroughly predictable pattern. Offspring having green seeds and those having yellow seeds both appeared in succeeding generations according to an unchanging mathematical ratio.

Mendel found that crossing pure-yellow seeds with pure-green seeds would inevitably result in a totally yellow-seeded first generation. But the green color would invariably return in the second generation in one out of every four offspring. This evidence led him to two conclusions. First, certain types of characteristics showed apparent dominance over others in a cross between differentiated parents (the yellow color, as the dominant characteristic, would therefore appear every time in the hybird first-generation offspring). Second, the green characteristic had not simply been wiped out during the crossbreeding, for it did reappear in the succeeding generations. The disappearance of the green color was thus not real; the two characteristics, green and yellow, had been transmitted to the offspring as separate elements, and each had retained its purity.

Mendel drew the scheme of the peas' color characteristics in this way:

YY   yy

Yy   Yy   Yy   Yy   (F_1-all yellow)

YY   Yy   yY   yy   (F_2- 3 yellow, 1 green)

Y - dominant characteristic (yellow)

y - recessive characteristic (green)

F - generation

In perceiving the characteristics as he did, with green (y) and yellow (Y) as the two elements of a single trait, Mendel had made the first concrete discovery of the gene.

While continuing his experiments, Mendel found that each succeeding generation behaved in precisely the same way as the first ones, with dominant genes manifesting themselves both in each pure dominant and each hybrid plant, and recessive genes showing up only in each pure recessive plant. And the same ratios existed, regardless of whether he worked with one or more characteristics at a time. Each and every characteristic revealed the same pattern as the one for seed color; each related pair of genes acted independently of the others to fulfill the mathematical predictions his observations had uncovered. After centuries, here was the breakthrough; the world of heredity was as logical and understandable as any other science, as accessible as simple geometry.

Mendel's experiments should have descended upon the infant world of biology like a bolt from heaven. But nothing of the sort happened. He was too far ahead of his time. His work was acknowledged by the Brno Society of Natural Science (and published in its journal) and then forgotten. Although he and his pea plants have given researchers who followed him a precise, quantitative experimental procedure for analyzing variations in genetic inheritance, his findings lay buried below the horizons of scientific awareness for over thirty years.

In the meantime, in the expanding world of the microscope, parallel discoveries were being made. By 1860 it was accepted that hereditary transmission from parent to offspring took place as a result of the union of sperm and egg; in 1868, Haeckel postulated that the site of heredity was in the nucleus; and in the same year, Friedrich Meischer uncovered the existence of DNA. Meischer, a sad-eyed, balding, bearded chemist, had been trying to discover the chemical composition of the cell nucleus. While working with pus cells and salmon sperm—cells with relatively large nuclei—he noticed that they contained a substance rich in phosphorus, which he called "nuclein." Ironically, although he was searching for the physical mechanism of cell reproduction in the nucleus, he never realized that his nuclein, later called nucleic acid, was the key.

The discoveries of Mendel and Meischer took place only three years apart. Yet it was more than thirty years before Mendel's work was "rediscovered" and the two were tied together. These two archetypal projects typify the divisions that have existed in biological research over the years. Mendel's work was phenotypic—he examined the questions of heredity by delving into its visible manifestations. Meischer, on the other hand, focused on the contents of the cell, an approach based on the belief that, since function determines structure, the most direct route to finding out how something works is through discovering how it is constructed. These two divergent approaches attest to the myriad ways in which any single scientific problem can be attacked. They also support the axiom that prediction without evidence is folly. Estimating the practical impact of a proposed course of basic research is like guessing at the number of angels gathered on the head of a pin; even the best estimate is hardly supportable.

It was not until the turn of the century that significant new parts of the hereditary puzzle began to fall into place. Albrecht Kossel, who had isolated ribonucleic acid (RNA) in 1892, identified the four bases, or nucleotides, of the nucleic acids as two purines—adenine and guanine—and two pyrimidines—cytosine and uracil—from a study of yeast and thymus. The discovery seemed of little significance, for Kossel believed that proteins, far more active and complex than nucleic acid, were the building blocks upon which life and heredity were based. He acknowledged that DNA was in some way connected with the synthesis of fresh tissue. But he—and those who followed—accorded the nucleic acids a secondary role in the transfer of heredity, a role which, because of their relatively simple structure, seemed entirely plausible. After all, who could expect that a molecule composed of little more than four bases, some sugar, and a bit of phosphorus could generate the intricacies and complexities of life that confront us every day? It did seem absurd.

Whether they had set their theories on the right track or not, biologists had begun to extract the facts from the mysteries of heredity as if they were peeling layers from an onion. William Bateson, who coined the term "genetics" in his book *A Defense of Mendel's Principles of Heredity*, acknowledged the "rediscovery" of the monk's findings and noted that "an exact determination of the laws of heredity will probably work more change in man's outlook on the world, and in his power over nature, than any other advance in natural knowledge that can be clearly foreseen." Today, with the imminence of genetic engineering, the mark of destiny that Bateson attached to the understanding of the gene seems to be coming true.

With the early 1900s came the discovery of the small, peripheral kinds of information so crucial to the understanding of any complex concept. Archibald Garrod recognized the apparent relationship between enzymes and genes—he observed that genes probably produced enzymes—in his study of alkaptonuria, an arthritic condition characterized by wine-colored urine. Garrod found that if someone with alkaptonuria mated with someone whose family history showed no indication of the disease, it would manifest itself in future generations exactly as Mendel's generational scheme had predicted. It became evident to him that alkaptonuria was a genetic disease, caused by the conjoining of two recessive genes. In addition, he noted that alkaptonuria occurred only in conjunction with the failure of one of the body's normal enzymatic reactions. From this, he theorized that the recessive gene responsible for alkaptonuria was a defective gene which failed to produce the very enzyme that would prevent the disease. Garrod's work constituted one of the first inquiries into the chemical nature of heredity. It established the direct link between genes and chemical reactions, and presaged the work that would ultimately lead to the identification and characterization of DNA as the genetic material.

By the 1920s, T.H. Morgan had begun mapping the chromosomes of fruit flies to determine the locations of the determinants of various characteristics; P.A. Levene had discovered the difference between DNA and RNA (the former was composed of the five-carbon sugar deoxyribose; the latter was composed of ribose); and Levene had also noted that thymine—and not uracil—was the fourth base in DNA's structure. But despite these discoveries, still nobody understood DNA's biological role. Its composition had been uncovered; its location within the nucleus had been confirmed; its connection with some part of the genetic process was assumed; but its function and actual structure remained out of reach.

To fill the theoretical chasm, Levene hypothesized that DNA was relatively unimportant to the replication of the cell. His tetranucleotide (four-nucleotide) hypothesis suggested that the four bases which made up DNA were all present in equal parts. If this were so, it would be impossible for DNA to provide the biochemical individuality that heredity demanded. There was simply not enough variation to account for it. Levene's theory gained acceptance and held sway until the 1940s. It lessened scientific interest in DNA and focused attention on proteins, which were known to be both large and varied, and were obviously complex enough to account for any number of genetic possibilities. In addition, proteins existed, with DNA, along the chromosomes. Thus their location as well as their composition seemed to point to their genetic importance. They, too, had a connection with enzymes; they were composed of amino acids, the fundamental evolutionary structures without which life cannot exist; they were richly present in all cells. Proteins were the obvious and logical choice for the hereditary material. As for DNA, Levene assumed that it performed an important role in the formation of proteins, acting as a kind of midwife, supervising their construction. To perform this function, DNA did not have to be specific. It did not have to, in any way, decide which proteins would be formed. It needed only to act mechanically, as a part of a production line.

The big problem with the tetranucleotide theory and the assumption that proteins were the building blocks of life was the fundamental question of cell replication and its place in heredity. Nobody yet knew how genetic material duplicated itself. If, for example, enzymes (which are proteins) created other enzymes (other proteins) with the help of DNA, where was the first enzyme created? As the progression looked, it seemed unlikely that there was some as yet undiscovered prime enzyme which could create itself out of nothing. No, a way had to exist for the genetic material to replicate itself, to begin the process of heredity without outside assistance. Otherwise, the entire protein theory stood on shaky legs.

By the first years of the Second World War, biologists had thus come up with a theory that they could support. Yet many of them realized that its tenets were like pieces of a jigsaw puzzle that would not quite fit together. New discoveries had added detail to the picture: a link had been discovered between the genetic map of a cell and the distribution of nucleic acid, and the one-gene/one-enzyme theory—which postulated that each gene produced a single enzyme to begin the chain of life—had been proposed. But biologists still had no idea what a gene actually was, although they bandied the term around quite naturally. To them, it still existed as Mendel had envisioned it eighty years before—an indivisible, amorphous, mysterious unit of heredity. And nobody had even begun to take apart the chemical reactions that caused the transmission of inherited characteristics, much less to attempt to discover which structures within their treasured proteins might make heredity possible. The pieces simply did not fit; but the picture they formed was so inviting that the scientists were not above forcing them together in the hope that their beauty implied logic as well. As Robert Olby has noted, "All that could be said at the close of the first three decades of Mendelian research was that many well known . . . genes controlled character differences whose basis was the presence or absence of a specific chemical reaction, this reaction presumably having been brought about by an enzyme, which in turn was the product of the gene, unless of course gene and enzyme were one and the same thing."[1] The science of genetics was still vague, general, and contradictory. But researchers had pushed their ignorance to the very edge of important discoveries. The next ten years would tell the tale.

Avery, McCarty, and MacLeod

When the first breakthrough did come, it emerged from an unexpected source, rather than from the indirect approach of the Mendelians or the analytical attempts of the microscopists. And it came in a manner so unlikely that its discoverers were as surprised as the rest of the scientific community at their results. The answer seemed to have appeared, almost magically, out of nowhere.

Oswald Avery, Maclyn McCarty, and Colin MacLeod had been working at the Rockefeller Institute in New York City, trying to isolate the agent that causes bacteria to mutate spontaneously from one form to another. It had been known since the 1920s that pneumococcus, the bacterium responsible for pneumonia, existed in at least four different forms, each of which could give rise to any of the others. In 1943, Luria and Delbruck had found that this transformation, or mutation, was spontaneous and required no external stimuli. Avery and his group were aware of these results as they began their search for the agent behind these spontaneous transformations.

The easiest way to prove that an agent is responsible for an action is to remove it from the cell and determine whether the action still occurs. If it does, the agent removed is obviously not the cause of the action; if it does not, the odds are that the agent is involved. Because he did not have the techniques to actually remove various agents from pneumococcus, Avery took this concept one step further. He could achieve the same results by destroying each

[1] Robert Olby, *The Path to the Double Helix* (Seattle: University of Washington Press, 1974), p. 123.

individual agent's ability to function, by inserting into the
medium in which the bacteria lived a destructive enzyme
which acted upon the agent he wished to test. Since enzymes only act upon substances for which they are specifically coded, he could discover the transforming agent by
determining which enzyme prevented the transformation
from occurring.

Because the group believed in the validity of the tetranucleotide hypothesis and assumed that proteins were
the probable constituents of genetic material, the first set
of enzymes Avery used were designed to destroy the function of proteins alone. They failed to interrupt the transformation process, and so he turned to RNase, the enzyme
responsible for breaking down the nucleic acid RNA.
When that, too, failed to work, Avery took P.A. Levene's
enzyme, DNase, and introduced it into the bacteria's environment. With DNase present, the transforming agent was
deactivated.

These initial experiments suggested a variety of possibilities. If, for example, it *was* DNA that controlled the
transformation of the bacteria, perhaps it was also DNA
which determined the original structure of the bacteria—
perhaps DNA, and not protein, was the true genetic
material. To test this hypothesis, the group added carefully
prepared DNA from one of the forms of pneumococcus to
another. The results of this experiment were clear; the hereditary properties of the second form were specifically altered by the addition of the first form's DNA.

With this experiment, Avery, McCarty, and MacLeod
hammered the first nail into the coffin of the protein theory of genetics. Although their published findings dodged
the issue of whether a link existed between biochemical individuality and DNA, the implications were quite clear.
And while Avery steadfastly held to the narrower view,
defending his research while staying away from public
speculation, few were willing to contradict his findings.
Isolated papers did claim that the evidence was by no
means conclusive, that Avery's research was tainted by im-
pure DNA, and that the project itself, directed as it was
by medical doctors rather than by research scientists, was
somehow innately inaccurate. But independent supporting
evidence began to surface. Others reported that crystalline

DNase, the pure enzyme, did indeed prevent the transformation of bacteria; experiments showed that Avery's DNA could have been contaminated with, at most, 0.02 percent of protein under the experimental conditions that he described; and in 1945, Boivin, a Frenchman, independently confirmed Avery's work by isolating the transforming agent, DNA, in the bacterium *E. coli*. When Hershey and Chase, working seven years later in the Cold Spring Harbor laboratory on Long Island, discovered that viruses—which consist of little more than DNA surrounded by a protein coat—infect bacteria with their DNA, using their protein only as an instrument to break through the cell wall, little doubt was left as to the genetic significance of DNA.

Avery's work effectively destroyed the remnants of the Mendelians' belief in the indivisible gene by establishing a chemical basis for the actions of enzymes and bacterial transformation. It also triggered a serious search for the chemical structure of nucleic acids by suggesting that there were wide variations in the way DNA was put together. As the focus on DNA sharpened, scientists began to understand the logical progression of their efforts. Increasingly, DNA was being accepted as the one and only source of genetic information. Now someone had to determine exactly how the four bases in DNA related to each other.

Erwin Chargaff and Base Ratios

The end of the tetranucleotide hypothesis finally arrived in 1949, when Erwin Chargaff published his findings on the relationships between the bases of DNA.

Chargaff's approach to the problem was to take DNA from various sources—from calf thymus, beef spleen, yeast, and tubercle bacillus (the agent of tuberculosis)—and measure the relative amounts of the four bases (adenine, guanine, cytosine, and thymine) in each one. If the tetranucleotide theory were indeed true, the four bases would appear in equal proportions in the various types of DNA; if it were not, perhaps some other ratio would become apparent. What Chargaff found was startling and simple. From the four bases, two sets of pairs emerged, and their ratios were constant. Regardless of the source of the DNA, the amount of guanine equaled the amount of cytosine; the amount of adenine equaled the amount of thymine. These 1:1 ratios were not random; they remained constant despite the acknowledged differences among the species tested; but the ratios of the guanine-cytosine pair to the adenine-thymine pair bore no similarities among species. They changed radically for every comparison that Chargaff could dream up. He therefore constructed an equation that summarized his findings, an equation that was to have a profound effect upon the future of genetics:

$$\frac{\text{Guanine}}{\text{Cytosine}} = \frac{\text{Adenine}}{\text{Thymine}} = 1$$

The bases adenine and thymine are both known as purines, while guanine and cytosine are both pyrimidines. The significance of this distinction would not be realized until the secrets of DNA's structure would at last be unraveled several years later.

And so the tetranucleotide theory of P.A. Levene was finally laid to rest. Chargaff's base ratios implied that the DNA molecule could be as complex as it had to be to fulfill the myriad needs of heredity; the only restriction in the details of its code would lie in the limits of the DNA molecule itself.

The constituents of the DNA molecule—and how they related to each other—had now been surmised. With the concept of base ratios came the opportunity for a final breakthrough—an assault on the molecule itself, and the mapping of its structure.

Crick, Watson, Franklin, Wilkins, and the Structure of DNA

The story of Watson and Crick's efforts to discover the secrets behind DNA's structure is one of the great dramas of modern science. The three major books on the subject—James Watson's self-laudatory *The Double Helix,*

Anne Sayre's revisionist *Rosalind Franklin and DNA*, and Robert Olby's painstakingly detailed *The Path to the Double Helix*—unfold a tale of the extraordinary collaboration of two men, of fantastic flights of imagination and scientific curiosity, of intrigue, of a race for a prize that could strip the chains from the fields of molecular biology and genetics. The characters themselves positively bled conflict. There was James Watson, an American on a postdoctoral fellowship who had received his doctorate in biology at the age of twenty, a bright, arrogant scrapper with little practical background in genetics who had come to the Cavendish Laboratory at Cambridge University to work on a project only remotely related to uncovering the structure of DNA. He collaborated with Francis Crick, his opposite in personality and drive, a man with awesome perceptions and a profound and broad brilliance that enabled him to meddle constructively in almost any project that came along, yet a scholar so undisciplined that, at the age of thirty-five, he had yet to complete his doctoral thesis. At nearby King's College in London was Rosalind Franklin, assigned to work on the DNA project, a woman in a field occupied almost exclusively by men. Franklin had a personality so assertive and a mind so quick that she spent hours of her time and energy in constant bitter conflict with those, like her nominal superior, Maurice Wilkins, who thought she would have done better in a kitchen with three children pulling at her apron. And finally, almost 9,-000 miles away at the California Institute of Technology was Linus Pauling, already a world-renowned scientist with credits a mile long and a reputation as one of the finest minds alive. To some, like Wilkins and Franklin, the task of finding the structure was a question of science and the pleasure of discovery itself; to others, like Pauling and Crick, it all seemed like a race; but to those like James Watson, the competition was closer to war. Nobody could be certain what the impact of the discovery of the structure of DNA would be. But the odds were that its effect upon the future of genetic research would be as important as the discovery of nuclear fission was to physics. The goal was obvious, the pot of gold at the end could be taken for granted; the race was on.

Crick and Watson began their collaboration in 1951.

Partly because each was involved in other projects as well, it took them almost two years—until February 1953—to come up with the correct solution. In between were the failures, as model after model of DNA fell before the data already assembled. Despite all that had been discovered, trying to map the structure of the molecule turned out to be slightly more difficult than being handed all the metal, plastic, glass, and tools necessary to create a car and then being asked to build a functional vehicle—without ever having seen one. Crick and Watson's solution was to use whatever hard structural evidence they could find to build theoretical models, and then to demonstrate that their guess conformed to the rest of the known information. A similar technique had been employed by Linus Pauling several years before when he worked out one of the secondary structures of a protein molecule, the alpha-helix, a monumental discovery which brought the helical shape into fashion and presaged the ultimate discovery of DNA's structure.

The problem in itself was mainly theoretical; there was very little physical evidence to go on. And the number of variables that had to be considered in the solution was astounding. There was the problem of shape: was the molecule round, irregular, or a helix—with one, two, or three chains of backbones intertwined? The phosphates and sugars were the components of the backbone—that was assumed—but were they on the inside of the molecule, or on the outside? What atoms bonded the bases to each other? What were the precise angles that existed between atoms? In what kind of order did the bases join, if they joined at all? The list seemed endless, but it included the one question which would decide whether the discovery would shake the biological world or would be recognized as merely important: did the structure of DNA provide a means for self-replication—for dividing and producing another identical strand? If DNA was, indeed, the genetic material, it almost had to have such a mechanism, for replication required that the cells divide and recreate an exact copy of the parent cell in each new generation.

By December 1952, Crick and Watson seemed well on their way to solving the puzzle. They had talked to Chargaff and had learned from him about the 1:1 ratio of his

base pairs; they had agreed that the molecule had a helical
shape; they had built and discarded three-stranded helical
models and were becoming convinced of the validity of a
two-stranded model; and, from Rosalind Franklin's published work on the results of her crystallographic photo-
graphs of light diffracted from a DNA molecule, they had
learned crucial information about the molecule's water
content and the distance between bases.

Then they learned through Linus Pauling's son that
Pauling was about to propose a model for the structure of
DNA. They were crushed. The race seemed over. There
was little chance that someone of Linus Pauling's stature
could have made a mistake. Their efforts had been wasted.

And then Watson read a copy of the article announcing
the discovery that Pauling and his collaborator were going
to publish in the *Proceedings of the National Academy of
Science*. He looked at it—and looked again. Incredibly,
Pauling had made an elementary error, an error which
made his structure a physical impossibility. The true structure was still a mystery, and now Watson and Crick had a
small head start—the six weeks before Pauling's article
would be published and his error recognized. They got to
work.

In January 1953, Watson visited the King's College laboratories of Wilkins and Franklin. Relations between Watson and Wilkins were warm, but between Franklin and
Watson there existed several sources of friction. First and
foremost, Watson and Crick's research into the structure
of DNA was in violation of an oral agreement between
Cavendish and King's College that placed the structure of
DNA in the hands of King's. Crick and Watson had, in
fact, been ordered by their superior, Sir Lawrence Bragg,
to stop their work on DNA the year before. Second,
Franklin disapproved of the methods that the two collabo-
rators were using in their work. Franklin believed in pure
science; she wanted to solve the structure by direct empirical methods, using her skill in x-ray crystallography (a
method of photographing the diffraction of light passing
through a molecule) to give her the measurements and dimensions necessary for solving the puzzle. She thought
little of model-building and the assumptions that working
with hypothetical structures generated. Crick and Watson

seemed like interlopers to her. Without having performed any of the basic work required for accurately determining the measurable elements of DNA, they were using her calculations and those of others in what seemed to be a fruitless, frivolous search. Finally, it was evident to her that Watson sympathized with Wilkins in the vicious feud that had erupted between the two King's College colleagues.

Wilkins and Watson talked privately. In the course of the conversation, Watson began to realize that his and Crick's assumption that DNA was a double helix was probably correct. Watson went back home to build several more faulty models, but returned to King's College for another visit in early February, a visit that later became far more significant than the information it managed to produce. For it was during this second visit that Wilkins showed Watson, without authorization, one of Rosalind Franklin's better unpublished crystallographic photos of DNA. Wilkins was trying to make a point about Franklin's refusal to recognize several characteristics of DNA that were obvious even in her own pictures. But Watson found something else in the photograph; from it he was able to discern the answers to some crucial problems that had been hindering his and Crick's progress. With this new information locked firmly in his mind, he returned to Cambridge.

Back in his laboratory, Watson, with Crick's guidance, began to build a new model, placing the sugar-phosphate backbones in two strands on the outside of the molecule and pairing the bases in the middle. But this model was destined for failure too, for, while he had finally decided upon the importance of base pairing, he had hit upon the wrong scheme. Instead of matching like with unlike bases, he tried to pair like with like. And although the result was a workable model, the structure fell apart under the scrutiny of Crick and an American crystallographer, Jerry Donohue.

Finally, during the ensuing discussion, the three researchers recognized the possibility of pairing a purine with a pyrimidine, of joining two unlike bases. And on the last day of February 1953, Watson sat down and began shifting the four bases around, trying the various possibilities:

. . . Suddenly I became aware that an adenine-thymine pair held together by two hydrogen bonds was identical in shape to a guanine-cytosine pair held together by at least two hydrogen bonds. All the hydrogen bonds seemed to form naturally; no fudging was required to make the two types of base pairs identical in shape. Quickly I called Jerry over to ask him whether this time he had any objection to my new base pairs.

When he said no, my morale skyrocketed, for I suspected that we now had the answer to the riddle of why the number of purine residues exactly equaled the number of pyrimidine residues. Two irregular sequences of bases could be regularly packed in the center of a helix if a purine always hydrogen-bonded to a pyrimidine. Furthermore, the hydrogen-bonding requirement meant that adenine would always pair with thymine, while guanine could pair only with cytosine. Chargaff's rules then suddenly stood out as a consequence of a double-helical structure for DNA. Even more exciting, this type of double helix suggested a replication scheme much more satisfactory than my . . . like-with-like pairing. Always pairing adenine with thymine and guanine with cytosine meant that the base sequences of the two intertwined chains were complementary to each other. Given the base sequence of one chain, that of its partner was automatically determined. Conceptually, it was thus very easy to visualize how a single chain could be the template for the synthesis of a chain with the complementary sequence.[2]

And there it was. Each strand of DNA consisted of a chain of nucleotides, with each nucleotide composed of a deoxyribose sugar ring, a phosphate group, and one of the four bases. The sugars and phosphates formed, together, an external backbone from which the bases projected. When two single chains of nucleotides were interwoven and the complementary bases were joined, the final, complete molecule took on a double-stranded, helical shape, a spiral ladder in which the paired bases acted as rungs. (Fig. 1).

Crick's reaction to the new model was as enthusiastic as Watson's. This new solution was as natural as any that they had previously devised. It fulfilled every criterion they could propose. It was simple and elegant, beautiful in its

[2] James D. Watson, *The Double Helix* (New York: New American Library, 1968), pp. 123-5.

FIG. 1. *The double-helical structure of DNA.*

symmetry and in the way it solved problems that had been plaguing scientists for years. And, most important, it provided a definitive answer to the question of self-replication, an answer which would effectively end any debate as to DNA's dominance of the genetic totem pole, while it provided an almost revolutionary insight into the most basic workings of nature. Although they had discovered neither how the chains of the double-stranded helix divided for duplication nor how each single strand recognized the correct base pair when it reformed, it was obvious that "If the actual order of the bases on one of the pair of chains were given, one could write down the exact order of the bases on the other one, because of the specific pairing. Thus one chain is, as it were, the complement of the other, and it is this feature which suggests how the desoxyribonucleic acid molecule might duplicate itself."[3]

Prophesies about the importance of the discovery came true, and in 1962, Watson, Crick, and Wilkins shared the Nobel Prize for their work in unraveling the structure of DNA. Rosalind Franklin was not so honored; she had died of cancer five years before. The irony, for her, was particularly touching. She never really knew from where Watson had gleaned the critical information which permitted him and Crick to devise the final solution so quickly. And when they succeeded, she herself was just a few steps from the truth. When asked how close he thought Franklin was to solving the puzzle herself, Francis Crick said that he supposed she was perhaps six weeks, perhaps three months from the answer. But glory and honor traditionally go to the victors; second place is no place at all.

The ceremony in Stockholm was not the only outcome of the Crick-Watson breakthrough; it also marked one of

[3] James D. Watson and Francis Crick, "Genetical Implications of the Structure of Desoxyribose Nucleic Acid," *Nature*, 171, 737-8 (1953); quoted in Olby, p. 427.

the earliest and most blatant manifestations of what has come to be known as the "Nobel Prize syndrome," a kind of neurosis characterized by unseemly haste which has overcome legions of otherwise reasonable men, pushing them toward acts of questionable safety and ethics for the sake of the reward. Although the problem seldom affects those outside the scientific community, a set of unique circumstances—for example, research in which an accident may lead to destruction beyond the confines of the laboratory—can vastly increase the harmful effects of the syndrome. It is this question of motive that has hounded the scientists who want the continuation of relatively unfettered research with recombinant DNA, and that has led opponents of the research to ask whether its proponents actually are arguing from a clear understanding of the facts or from a position of self-interest. It is a charge that will continue to color the debate on the relationship between science and society for decades.

The Road to Recombination

While the knowledge of DNA's structure made it clear how the molecule could replicate itself, neither the mechanisms of replication nor the way it was catalyzed had been uncovered. Determining the structure of DNA was an achievement. But it was merely one step in a series necessary to take the scientific knowledge of heredity out of the laboratory and put it into practice. The goal of the research, beyond scientific discovery, was to be able to manipulate the functional units of DNA—the genes—a skill which could open the door to control of the hereditary process.

From the beginning, enzymes seemed to be the key to how DNA replicated itself and passed on genetic information. Enzymes are proteins. They act as catalysts for most biochemical processes, triggering the chemical transformations for which many substances, including other proteins, have been constructed. Enzymes were quickly found to be responsible both for the division of the DNA molecule into two complementary templates, and for each single strand's recognition of its own correct base-pairing scheme. And while important work on DNA advanced over a broad front—the 1960s saw the demonstration that complementary DNA molecules could be separated and recombined, and Francis Crick and his colleagues discovered that the key to the genetic language lay hidden within the sequence of the bases in the molecule—some of the most important discoveries concerned the isolation and identification of relevant enzymes.

In 1967, DNA ligases—enzymes which can repair breaks in DNA—were discovered simultaneously in five laboratories. When a break between two nucleotides occurs, DNA ligase catalyzes the synthesis of a bond precisely at the site of the break, reestablishing the link between the sugar of one nucleotide and the phosphate of its neighbor. The isolation of DNA ligase gave researchers a powerful tool with which to put disintegrated DNA back together again. It was also a first step in the specific process which would ultimately make it possible to recombine two different types of DNA. Indeed, by 1970, H. Gobind Khorana, then working at the University of Wisconsin, found that one type of ligase could actually catalyze the end-to-end linkage of separate segments of DNA if the two segments were kept close while the ligase acted. Because the positioning of the segments was a matter of chance, and because no method had been found to bring them together, the reaction was terribly inefficient. Still, it was clear that, if a mechanism could be developed that would hold the two DNA ends together, the process had enormous potential.

Shortly thereafter, a complex method for achieving this was devised in two laboratories at Stanford University. And in one of the laboratories, Paul Berg, David Jackson, and Robert Symons recognized the potential of joining strands of DNA from different organisms at will and planned the first experiment in the artificial, directed chemical recombination of DNA.

Berg began his work with DNA to learn something about the regulatory mechanisms that function within animal cells to control their reproduction. Early on, he saw that one way of exploring the molecular biology of gene expression in higher organisms was to use, as an analogy, a far simpler system: the system provided by the small viruses, the microorganisms that consist of little more than a squiggle of DNA surrounded by a protein coat. Since the group of viruses he had decided to use—the bacteriophages—introduce their own DNA into bacteria in nature in order to reproduce, Berg and his colleagues decided to use their existing mechanisms to transfer (or transduce) genes from one bacterial cell to another. Berg's problem was to find a way to recombine the DNA he wanted to transfer to the new host with the DNA of the bacteriophage. If the bacteriophage would accept the foreign DNA and carry it into the host bacterium with its own, Berg could study the effects of the new composite DNA on its host simply by comparing its processes to those of similar bacteria that had not been modified. Any differences would be caused by the new DNA. If this concept actually worked, it would be relatively easy to characterize the actions of any gene that could be successfully implanted in a new host.

Because there is no naturally occurring mechanism that permits bacteriophages to recombine with another organism's DNA before carrying the hybrid DNA to a host, Berg and his group devised an artificial system. The ability of DNA ligase to repair breaks in a strand of DNA had been proved, as had Khorana's finding that two separate strands of DNA could, indeed, be joined. And the Stanford group knew of another enzyme, terminal transferase, which (as its name implies) could catalyze the transfer of a series of identical bases to the end of a strand of DNA. The Crick-Watson double-helical model of DNA indicated that the molecule itself was composed of two intertwined strands. Each strand was made up of a chain of linked nucleotides. If terminal transferase could add nucleotides to one of the two strands of a complete DNA molecule as easily as it could to the end of a single strand, Berg could take a complete molecule and literally attach a tail of nucleotides to it. If, for example, he attached a string of adenines (one of the four bases) to the end of one molecule of DNA, he could join that molecule to an entirely different one by attaching a similar tail to it, composed entirely of thymine. By the rules of base pairing established by Crick and Watson, the two tails would then be complementary; hydrogen bonds and DNA ligase could be used to anneal them, exactly as if a break in a single molecule were being repaired. And Berg would have a molecule of recombined DNA to insert into his bacteria. (Figs. 2 A-D).

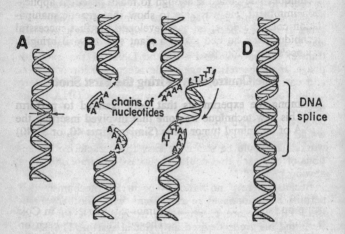

Fig. 2. *Paul Berg's first experiment in recombination.*
- (A) *The original DNA is divided by enzymatic action.*
- (B) *Chains of complementary nucleotides are added to the ends of both the original DNA and the foreign DNA.*
- (C) *The two strands of DNA join, linking up by means of their complementary bases.*
- (D) *Enzymes anneal the breaks in the DNA. The new, hybrid piece of DNA—a combination of original and foreign DNA—is ready to be inserted into a host bacterium.*

This chemical method for recombining DNA was extraordinarily delicate; each step required precise, painstaking methodology. Because the DNA used by Berg was actually circular, he and his group first had to divide it, using a special enzyme. Adding the nucleotides to the DNA molecule's ends was in itself a little like trying to thread a needle blindfolded. And once the "sticky ends" the extra nucleotides provided had been paired, it actually required three different enzymes to fill the gaps and seal the molecule.

Berg's work was not the first to imply a vision of genetic engineering, but it was the first intimation that the techniques had gone far enough to make practical application imminent. He was able to show that genetic manipulation could be done; he had developed the first successful technique; and he had taken a giant step toward bringing science's future into the present.

The Controversy: Firing the First Shot

Among the experiments that Berg planned to perform with his new technique was one that involved inserting the DNA of an animal tumor virus (Simian Virus 40, or SV40) into a bacteriophage (in this case, bacteriophage lambda) and then recombining this new hybrid with a host bacterium, *E. coli*. But in the summer of 1971, before Berg could actually perform the experiment, one of his graduate students, Janet Mertz, described the proposal during a workshop on laboratory safety at a tumor-virus meeting in Cold Spring Harbor, Long Island. The workshop had been organized by Robert Pollack, a Cold Spring Harbor biologist, precisely because of his fear that a laboratory accident with SV40 was not only possible but likely in the near future. When he heard Mertz' description of the experiments that Berg's group had planned, he was horrified.

Pollack had more than one reservation about the experiments being proposed. His first was with the use of SV40, a tumor virus commonly found in monkeys. Although SV40 is believed to be harmless to monkeys, it has been shown to cause cancer when injected into mice and hamsters. It also causes cellular changes when it is allowed to infect normal human cells in the laboratory. When SV40 was discovered in 1961, researchers realized that they had been unwittingly contaminating countless batches of polio vaccines with the virus in the years before; virtually millions of people had been given a dose of SV40 with the hypodermics and sugar cubes that contained the vaccine. Although SV40 has not been implicated in any human pathology resulting from that mass vaccination, several thousand people who were exposed to the virus are still being closely followed. Pollack's main concern over the dangers of SV40 had arisen primarily because his colleagues at Cold Spring Harbor were working on new ways to breed concentrated batches of the virus. And now, taking exception to his concern for the safety of both laboratory workers and the public in general was a researcher who, with few qualms, was proposing the transduction of SV40 genes into a fast-breeding, promiscuous host bacterium, *E. coli*.

Pollack's second major concern involved *E. coli* itself, a bacterium commonly found in the human gut. For decades, *E. coli* had been the bacterium of choice for research into biology and genetics. It had its advantages: more was known about *E. coli*'s processes than about any comparable bacterium; it divides at a rate of almost once every twenty minutes (providing a range of almost a hundred generations each day); and it easily exchanges genetic information with other *E. coli*. But these were the precise reasons that Pollack feared it. If the SV40 genes were inserted via the hybrid bacteriophage into *E. coli*, there was no way of telling whether the virus would transfer its code for activating cancer along with its other harmless properties. And if it did, and if the *E. coli* were to escape the laboratory, one natural new host for the dangerous hybrid would be man. "Janet Mertz talked of putting SV40 in [bacteriophage] lambda; lambda in *E. coli*; *E. coli* in something else," Pollack said. "But that something else was *people*."

Mertz had indeed considered the possibilities, but the dangers seemed to her to be remote. She felt that she was bringing up the subject not so much to counter Pollack's discussion of safety as to present a practical, imminent research topic for discussion. But she did not expect the reaction that her disclosure provoked.

In his agitation, Pollack called Paul Berg to ask him whether he could at least change the host, inserting hybrid DNA into some other bacterial host that does not make its home in the human gut. Berg was dumbfounded by Pollack's caution. He could see no dangers in what he was doing. Even as he thought about it, the risks seemed so remote as to be purely hypothetical. Would he actually be pressured into restricting legitimate scientific research because of some totally baseless fears? Nevertheless, he promised Pollack that he would talk to other researchers about the dangers, to establish some sort of consensus. Within six months, Berg called back. He had talked to a variety of people—from DNA researcher David Baltimore and Nobel laureate Joshua Lederberg to Daniel and Maxine Singer (he a lawyer, she a scientist working with the National Institutes of Health)—who had become involved in the questions of scientific ethics. The consensus was clear; although the probabilities of hazards actually occurring seemed low, there was no possible way of eliminating them. The risks did have some basis in scientific fact. Furthermore, Berg had realized that if his experiment with SV40 failed, there would be no payoff at all—the technique would simply go down the drain—whereas if he succeeded, he would be faced with the dangers of the hazards. The destructive potential of the experiments definitely outweighed the knowledge to be gained. He abandoned them.

During his discussions with his colleagues, Berg had floated the idea of holding a conference to discuss the growing issue of biological hazards and tumor-virus research. He therefore asked Pollack to help him organize a series of two scientific meetings which, he hoped, would carefully evaluate such risks. Pollack agreed to help. He was in accord with Berg's assessment that many of those experimenting with DNA were ignoring the problems of biohazards. The first conference could thus air the question and gather the pertinent facts; the second would discuss ongoing experiments, their hazards, and the state of present knowledge about the biohazards of tumor viruses.

The conferences would be held at the Asilomar Conference Center in Pacific Grove, California. Asilomar was a state park and retreat where scientific meetings had often been held in the past. It sits right on the ocean, in a redwood forest; there are few places more beautiful or more accessible that are also conducive to scientific discussion. Berg himself had held meetings there with his own group to provide an alternative to some of the normal tensions that build up in the confines of the laboratory.

Recombination: From Theory to Practice

While the organizers of the first Asilomar Conference were planning their meeting, other researchers were sustaining the momentum Paul Berg had let slip when he turned to the problems of biohazards. Much of it involved experimentation with *Eco R*I, the enzyme that Berg had used to cleave his original DNA during his first attempts at recombination.

*Eco R*I, called a restriction enzyme, is one of a group that performs a crucial task for bacteria in nature. Because foreign DNAs possess a variety of ways of entering bacteria naturally, bacteria have developed a method of protecting themselves against the dangers of the invasion; without this method, they would have no way of maintaining their own genetic integrity. By recognizing specific sequences of nucleotides on foreign DNA, and by cleaving the new DNA at those sites like tiny biological scissors— destroying its ability to function—restriction enzymes act as the bacteria's protective mechanism. Meanwhile, the bacteria protect themselves from the enzymes with chemical additions to their own DNA that make them resistant to the cutting process.

The existence of these restriction enzymes was discovered in 1962. In 1971, a group under Herbert Boyer, working at the University of California Medical Center in San Francisco, was able to isolate and purify *Eco R*I; the group also found that *Eco R*I leaves behind whole pieces of DNA large enough to contain from one to ten genes. By November 1972, Ronald Davis and Janet Mertz (whose confrontation with Robert Pollack triggered the debate on biohazards in the first place) reported that *Eco R*I naturally left the pieces of DNA with the very same types of "sticky ends" that Berg had been laboriously attaching to the ends of the molecule by means of the terminal transferase enzyme.

But the ability to cleave at precisely the same site on every molecule was not the only property of *Eco R*I that held the geneticists' interest. Two other properties were just as important. First, as just mentioned, *Eco R*I does not simply cut perpendicularly through the two strands of the DNA molecule, but on a bias, leaving the "sticky ends" discovered by Mertz and Davis at the end of each segment. Second, the particular site chosen by *Eco R*I to cleave is at a point of symmetry on the DNA molecule. That is, the nucleotides around the point of cleavage are perfectly balanced, in the way that palindromes, like "Otto," "563365," or "Poor Dan is in a droop" are bilaterally symmetrical, like mirror images.

In reality, the site of cleavage chosen by *Eco R*I looks like this:

Guanine—Adenine—Adenine—Thymine—Thymine—Cytosine
Cytosine—Thymine—Thymine—Adenine—Adenine—Guanine

Or, in genetic shorthand:

$$G—A—A—T—T—C$$
$$|\ \ |\ \ |\ \ |\ \ |\ \ |$$
$$C—T—T—A—A—G$$

*Eco R*I cuts the molecule between G and A on each of the two strands, leaving each end with a tail consisting of TTAA. Since any molecule cleaved by *Eco R*I is left with this same TTAA tail, and since the Crick-Watson scheme requires that base pairing be complementary, any molecule that *Eco R*I can cut can be recombined with any other through the interaction of the sticky ends. Thus, recombinant DNA molecules, containing parts of two different structures and joined by these chemical bonds, can be produced (Figs. 3, 4, 5, 6).

The impact of this discovery was tremendous. Instead of having to go through the time-consuming, painstaking process developed by Berg, all a researcher had to do to recombine two different types of DNA was to purify each by means of an ultracentrifuge, pour some *Eco R*I into each purified batch, and mix the two types together with a little DNA ligase (the repairing enzyme) to anneal the "sticky," complementary tails—the so-called "shotgun" method of experimentation. But the key to the research was not simply to recombine two types of DNA; recombination is a technique, a method of chemical bonding.

FOREIGN DNA

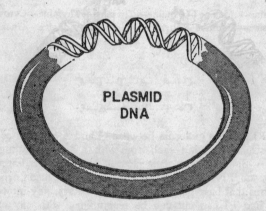

PLASMID DNA

FIG. 3. *Stanley Cohen's modified recombination. Foreign DNA is to be spliced into a plasmid.*

Unless a single, pure type of hybrid DNA could be isolated from the many in the test tube, and unless it could be made to successfully express itself in the bacteria into which it was being introduced, the procedure would become little more than an elegant, empty exercise.

In November 1972, at an international conference, Boyer met with Stanley Cohen, another researcher from Stanford, to discuss possible experiments that would lead to the successful expression of hybrid DNA. Instead of using the long, complex DNA contained in the chromosomes of *E. coli*, Boyer and Cohen decided to attempt their genetic manipulation with plasmids, the small, circular, secondary bits of DNA which float around bacteria independent of the primary genetic material. The plasmid,

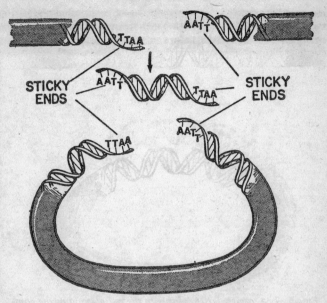

FIG. 4. *A restriction enzyme, Eco RI, cuts both pieces of DNA, leaving each with so-called "sticky ends."*

as a natural part of the bacterium, is already coded to replicate with it; therefore, any hybrid containing one entire functioning plasmid, plus the additional genes from another source, would have a good chance of replicating and expressing itself in the new host. Plasmids had another attractive feature. Because these free-floating bits of DNA are much smaller than the DNA chains contained in the chromosomes, they are much simpler and contain fewer genes. Boyer and Cohen needed a plasmid so simple that *Eco R*I would cleave it at only one site, leaving its chain of nucleotides intact. Then they could add the new genetic material at that site, anneal the two breaks with DNA ligase, and have a complete, circular hybrid plasmid, already coded to function within the bacteria.

Cohen's collection of plasmids at Stanford included one tiny plasmid, called plasmid Stanley Cohen 101 (*p*SC101),

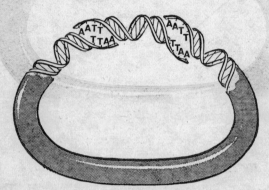

FIG. 5. *The foreign DNA joins with the cut plasmid as the complementary bases of the sticky ends attract.*

that had been isolated from a larger plasmid which bore
the genes for several types of antibiotic resistance. And
*p*SC101 was not only small and simple, but contained the
gene for one specific type of antibiotic resistance,
resistance to tetracycline. This meant that Cohen could actually test the success of his experiments simply by adding
tetracycline to the finished culture, for if he had begun his
experiment with *E. coli* sensitive to tetracycline, the only
bacteria that would survive the intrusion of tetracycline after the experiment would be those that had successfully
added the tetracycline-resistant *p*SC101 to their genetic
complement. This preliminary step succeeded: *E. coli* that
had been sensitive to tetracycline became resistant because
*p*SC101 had expressed itself in the bacteria.

The next task for Cohen and his assistant, Annie Chang,
was to create a hybrid *p*SC101 plasmid. They recombined
the plasmid DNA of an *E. coli*—which carried resistance

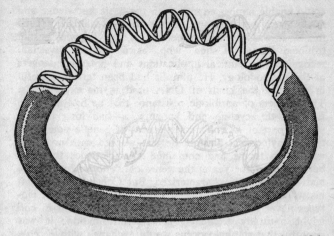

FIG. 6. *DNA ligase, an enzyme, anneals the breaks in the DNA, and a hybrid plasmid is formed. The new plasmid is ready to be inserted into a host bacterium.*

to a second antibiotic, kanamycin—with *p*SC101. Then they inserted this new hybrid plasmid into a fresh *E. coli* host. Again, some of the transformed bacteria became resistant to both tetracycline and kanamycin. The experiment was a complete success; the *p*SC101 plasmid, while expressing its own genetic content, had also proved capable of carrying with it a related segment of DNA which had been stripped of its own self-replicating mechanism.

Cohen and his group duplicated these experiments with recombination and antibiotic resistance, first with a plasmid from staphylococcus, a bacterium unrelated to *E. coli*, then with genes taken from the toad *Xenopus laevis*. Both experiments led to the successful expression of the foreign genes in *E. coli*.

By the summer of 1973, researchers had shown that genetic manipulation was not only possible but practical. Boyer and Cohen had pioneered a method of recombination conceptually different from Berg's first efforts, one which was far simpler and more direct, which could easily be applied to masses of bacteria all at once, and which could enable almost anyone with the basic knowledge and materials to perform any number of recombinant experiments. It was this last aspect of his new knowledge that troubled Stanley Cohen, who, as a trained physician, recognized the clinical implications and potential dangers of the methodology. He himself had been terribly careful in choosing the kinds of DNA used in his experiments. The markers of antibiotic resistance that he had added to *E. coli*—tetracycline and kanamycin—had long been a natural part of *E. coli*'s complement of genetic potentials; the staphylococcal plasmid he had worked with produced no known toxins and contained characteristics that had been mapped for years; the genes of *Xenopus laevis*, too, had been partially characterized. But Cohen realized that his work had the potential to be used to create novel biotypes never before found in nature. While related *E. coli* might naturally exchange genetic information, it was doubtful that staphylococcus would exchange with *E. coli*, and even more unlikely that any kind of animal gene would naturally recombine with the genetic complement of any bacteria, especially on the scale that was now possible. Biological recombination was not new; it had long been recognized as one element which provided for the normal genetic reassortment that was a part of evolution in nature. What was new was the opportunity for out-of-nature chemical transformation. Whereas the hazards of naturally transferred antibiotic resistance and genetic modification in disease-causing organisms had always been within the purview of nature, it was now possible to create combinations of hazardous genes that nature might not know how to handle.

In the midst of this growing awareness, the first Asilomar Conference was held. The attending scientists did little beyond show their lack of knowledge about the revolutionary techniques being devised. They demonstrated even less knowledge about the consequences. Although the conference did agree on one specific proposal—to monitor those working in laboratories experimenting with tumor viruses—both the idea of another conference and the monitoring proposal were eventually dropped. With little impetus for further discussion, the issue seemed destined to fade away. The media and public knew nothing of the potential controversy. Public scientific debate about genetic engineering revolved mainly around the ethics of applied research, the science-fiction-like questions about taking knowledge and using it to create hardware that can alter the human condition. For the moment at least, questions surrounding basic research—whether it in itself was dangerous; whether society should have control over an arm of science that might result in irreversible damage to the environment, the ecosystem, and even the genetic composition of man—seemed to have been put to an uneasy rest. Boyer and Cohen were continuing their research. Soon others were going to follow, finding vectors and vehicles other than *p*SC101, searching for newer, better hosts than *E. coli.*

But six months after the first Asilomar Conference, conferees at the Gordon Conference, catalyzed by Herbert Boyer's report on his successful experiments in recombination, decided to voice their concern, to raise the issue in public debate. Naively, almost blindly, they did not see where an issue of such potentially explosive content could lead.

TWO:

The Debate

The Gordon Conference

The Gordon Research Conference on Nucleic Acids was held in June 1973. During the conference, one of the papers presented was Herbert Boyer's discussion of the techniques he and Stanley Cohen had developed for simply and quickly recombining segments of DNA. At first his talk caused little debate. But gradually the participants at the conference began to realize the implications of Boyer's remarks. Maxine Singer—one of the scientists with whom Paul Berg had spoken after his telephone conversation with Robert Pollack—and Dieter Söll, the organizers of the conference, decided to revise the planned schedule, making time available for a discussion of the possible hazards of Boyer's work on the last day of the conference.

The discussion itself had little substance—it was clear that no experimental method yet existed to verify whether or not the research posed any real risks, nor did the conference have any basis for making quantitative judgments of hazards or benefits. What was evident was that the participants intended to at least voice their uneasiness. After

the early departure of many participants had depleted the
original 150 by about 40 percent, it was decided that a letter of "deep concern" should be sent to the National
Academy of Sciences and the National Institute of Medicine, two of the formal bodies monitoring the progress of
science in the United States. The conferees also agreed to
a second motion, which called for making the letter public, but by a margin of only three votes. Those who dis-
agreed felt that any media exposure would be premature
and counterproductive. They were leery of turning public
attention to something nobody knew anything about; no
matter what else happened, release of the letter to the
media would be impossible to undo. Their reluctance was
understandable. As businessmen, doctors, politicians, and
other professionals do, they were acting upon an instinc-
tive fear of the lynching mentality of an uninformed public. But if the second motion had been tabled, and if the
issue of recombination had been kept under wraps, the
damage to scientific credibility when the issue finally did
explode into the public consciousness would have been in-
calculable.

The letter, signed by Singer and Söll, was published in
Science magazine, one of the major U.S. scientific organs,
in September 1973. After briefly describing the new techniques, the letter succinctly expounded the conference's
concern:

. . . new kinds of hybrid plasmids or viruses, with biological
activity of unpredictable nature, may eventually be created.
These experiments offer exciting and interesting potential both
for advancing knowledge of fundamental biological processes
and for alleviation of human health problems.

Certain such hybrid molecules may prove hazardous to
laboratory workers and to the public. Although no hazard has
yet been established, prudence suggests that the potential hazard
be seriously considered.

. . . The conferees suggested that the Academies establish a
study committee to consider this problem and to recommend
specific actions or guidelines, should that seem appropriate.

The letter itself generated no great public stirring, and
its low-keyed tone caused the media to miss its signifi-

cance. But the scientific community did react. The National Academy of Science moved to set up an informal committee, led by Paul Berg, to gather the details of the potential biohazards.

The letter raised another question, one almost as important as the issue itself, an almost entirely new phenomenon which has dogged the debate ever since: Where *do* scientists involved in basic research go when they become concerned about some aspect of their work? To what body do they take their problems? Outside of the national academies and the various funding agencies, there exists almost nothing in the way of a formal organization to handle such situations.

The Gordon Conference thus became another step in the awakening of the scientific community to the issue of public safety. In reality, the conferees were only guessing at the dangers; the techniques discussed by Boyer were still in the formative stages. But their instincts were right. Through that summer and the next fall, Cohen and Chang completed the successful recombination of staphylococcus and *Xenopus laevis* with *E. coli.* They published their work in April 1974, just as the informal committee requested by NAS and chaired by Paul Berg met to first raise the possibility of a moratorium on research into recombinant DNA molecules.

The MIT Eight

Paul Berg had not been idle since NAS had contacted him. From his original sense of doubt, he had begun to emerge as the single most important voice moderating the growing controversy. From his status as one of the pioneers of the research, he had become one of its most influential overseers. As one interested party said, "[He] used judo on the whole argument, just flipped the whole thing over, and instead of being the guy who is the problem, he became the guy who is ostensibly the solution." Honest, dynamic, commanding, Berg had gained scientific prominence on the strength of his personality and willingness to work. He now focused that hard-won influence upon this new issue.

Casting about for ways in which to deal with the problem, Berg turned to James Watson, who, after his triumph with Crick in 1953, had held a professorship at Harvard before taking over direction of the important laboratory at Cold Spring Harbor. Watson had been a strong supporter of the first, unsuccessful Asilomar Conference in January 1973. He had also been an outspoken advocate of public safety during the debates over the wisdom of building nuclear reactors in the 1960s. Berg and Watson decided to call a meeting to examine the experiments being done and those that people still wanted to do, to critically evaluate whether they were safe (and what should be done if they were not), and to decide on a method for informing researchers of the rising concern for public safety that recombination was generating. To prepare for this larger meeting, they decided to hold a preliminary conference of their own group in April 1974 "to consider whether or not there is a *serious* problem growing out of present and projected experiments involving the construction of hybrid DNA molecules *in vitro* [in the test tube]. If a problem exists, then what can be done about it; both short and long-term actions. One of the Councils of the National Academy is seeking our guidance and hopefully, out of our meeting, we could offer some suggestions which would be helpful."

The question of safety had finally grown from a potential problem to a "serious" one. And Berg and Watson were not the only ones feeling the pressure. Stanley Cohen was being inundated by requests for supplies of his plasmid, *p*SC101. Although *p*SC101 is not the only plasmid that can make recombination work, it was then the only one whose efficacy had been proved. To either duplicate Cohen's research and recreate the tiny plasmid or find another plasmid which could act successfully as a vector for foreign genes would have taken months, if not years. Other researchers contemplating recombinant DNA research recognized a short cut when they saw it; they knew the value of beginning their experiments with something that had already proved successful.

Because they were concerned about the risks of the new research, Cohen and Chang had asked those requesting the plasmid to outline their proposed research, and they were appalled at some of the plans. Some proposed the very experiments that had triggered the debate in the first place: the insertion of tumor viruses into *E. coli*. Others wanted to perform recombinations both novel and frightening, like inserting herpes virus DNA into *E. coli*. Still others set forth plans that seemed little more than frivolous. As a result, Cohen and Chang devised a series of informal guidelines which researchers would have to meet if they were to receive the plasmid. But they were uncomfortable in their roles as judges of their colleagues' research. Berg's committee and its work would provide an outlet, a chance to share the responsibility, and a mechanism for dealing with the problem from the perspective of a group rather than an individual.

On April 17, 1974, eight scientists, all at the pinnacle of their profession, met at the Massachusetts Institute of Technology. The eight—Paul Berg (the chairman), David Baltimore, Herman Lewis, Daniel Nathans, Richard Roblin, James Watson, Sherman Weissman, and Norton Zinder—had been brought together at least partly because they all happened to be on the East Coast at the same time; a similar meeting on the West Coast would have consisted of an entirely different group, and might have led to an entirely different outcome.

The eight settled around a conference table with no firm agenda, and spent the first few hours in an informal bull session, playing the time-honored game of one-upmanship, with each trying to imagine a scenario of danger in recombinant research more horrifying than the last.

When they settled down, the scientists began to examine the possible approaches to the problem. At first there was considerable uncertainty about how to proceed. Although everyone agreed that a conference might be necessary, nobody knew how to handle the period of time between their meeting and the conference itself. Finally, Norton Zinder suggested the idea of a moratorium on those types of research that seemed to contain the greatest potential for risk. Similar concepts had been rolling around in the minds of Baltimore and Roblin, and the group discussed how to go about requesting such a moratorium (which, because of the committee's unofficial, informal nature, would be completely voluntary) and which experiments should be covered. The greatest drawback stemmed from the scientists' philosophical predilection for unfettered scientific research; any type of restriction, whether or not it was voluntary, seemed to set a precedent that might inevitably impede the flow of future progress. But looming larger than their personal preferences were the practical problems of potential risks, even though there was, as yet, no evidence that they existed. Their reluctance to initiate a moratorium was both understandable and ironic; science has accepted so many external restrictions on scope and range of research—from regulations on human experimentation and nuclear research to the informal, yet powerful, boundaries drawn by the funding agencies who control the purse strings and, *de facto,* the direction of research—that the proposed moratorium was not nearly as great a break with tradition as it seemed. Nevertheless the doubts persisted.

The Letter

Richard Roblin was appointed to write the first draft of a letter—which, the scientists recognized, would have to be made public. Roblin went home to New York and began working. His draft dealt with the two types of research that caused the most concern: the recombination of genes carrying antibiotic resistance, and recombination with oncogenic—cancer-causing—viruses. Although the ability of bacteria to gain genetic resistance to certain antibiotics in nature was well known, the group feared that researchers might take an antibiotic to which *E. coli,* for example, was still sensitive, give the bacteria the ability to resist, permit the newly formed resistant bacteria to escape, and thus destroy the medical effectiveness of the antibiotic. The theoretical dangers inherent in recombining *E. coli* with oncogenic viruses were obvious.

To these two types of research, Roblin added a third category. Because certain types of animal DNA contain sequences of nucleotides similar to RNA tumor viruses, he decided to add a proviso requesting extreme care in work with animal-cell DNA. The group had split on the issue of a moratorium for this kind of research. Roblin's compromise avoided the moratorium, but clarified the potential dangers of the research and placed the onus of safety upon the researcher.

The letter went through several drafts. By the third draft, it had reached its final form. Berg showed it to several researchers in California—Stanley Cohen, Herbert Boyer, Ronald Davis, and David Hogness—who requested that their signatures be added, partly because, of the original signers, only Berg had had direct experience with recombinant research. The third draft was sent to the National Academy of Sciences, which had convened the Berg group in the first place and now gave it official status as the academy's Committee on Recombinant DNA Molecules. After the NAS made slight modifications, the letter was readied for publication on July 18, 1974.

The final form of the letter began with a short outline of the research that had been accomplished and a small sampling of the dangers that might arise. Then the committee stated its recommendations:

First, and most important, that until the potential hazards of such recombinant DNA molecules have been better evaluated or until adequate methods are developed for preventing their spread, scientists throughout the world join with the members of this committee in voluntarily deferring the following types of experiments.

Type 1: Construction of new, autonomously replicating bacterial plasmids that might result in the introduction of genetic determinants for antibiotic resistance or bacterial toxin formation into bacterial strains that do not at present carry such determinants; or construction of new bacterial plasmids containing combinations of resistance to clinically useful antibiotics unless plasmids containing such combinations . . . already exist in nature.

Type 2: Linkage of all or segments of the DNA's from oncogenic or other animal viruses to autonomously replicating DNA elements such as bacterial plasmids [e.g. *p*SC101] or other viral DNA's. Such recombinant DNA molecules might be more easily disseminated to bacterial populations in humans and other species, and thus possibly increase the incidence of cancer or other diseases.

Second, plans to link fragments of animal DNA's to bacterial plasmid DNA or bacteriophage DNA should be carefully weighed in light of the fact that many types of animal cell

DNA's contain sequences common to RNA tumor viruses. Since joining of any foreign DNA to a DNA replication system creates new recombinant DNA molecules whose biological properties cannot be predicted with certainty, such experiments should not be undertaken lightly.

Third, the director of the National Institutes of Health is requested to give immediate consideration to establishing an advisory committee charged with (i) overseeing an experimental program to evaluate the potential biological and ecological hazards of the above types of recombinant DNA molecules; (ii) developing procedures which will minimize the spread of such molecules within human and other populations; and (iii) devising guidelines to be followed by investigators working with potentially hazardous recombinant DNA molecules.

Fourth, an international meeting of involved scientists from all over the world should be convened early in the coming year to review scientific progress in this area and to further discuss appropriate ways to deal with the potential biohazards of recombinant DNA molecules.

The letter recognized the difficulty of evaluating the research for hypothetical hazards. Yet it concluded: "Nonetheless, our concern for the possible unfortunate consequences of indiscriminate application of these techniques motivates us to urge all scientists working in this area to join us in agreeing not to initiate experiments of *types 1* and *2* above until attempts have been made to evaluate the hazards and some resolution of the outstanding questions has been achieved."

The letter was a model of decorum, tact, and caution. It recommended not the cessation of research in recombinant DNA but that others "join with the members of the committee in voluntarily deferring" two precisely defined types of experiments, while exercising the greatest care in conducting a third. It acknowledged the scientific community's ignorance of the potential dangers of the research and called for programs and meetings to provide the necessary information, yet carefully admitted that concern was "based on judgments of potential rather than demonstrated risk." This was not a letter written by scientists concerned with the misuse of genetics or its techniques. To them, the sole question concerned the potential risk of legitimate scientific experiments. When it was drafted, little thought was given to the reaction of the public or the media, for it was meant to be a letter from scientists to scientists. The members of the committee understood only a part of the difficulty of their task: they recognized that their call for a moratorium, or pause, would fundamentally change the way genetic research would be conducted. But their move did not seem unprecedented. Guidelines were available in other areas of potentially hazardous scientific activity—in work with radioactive materials, toxic chemicals, or naturally occurring disease-causing bacteria or viruses, for example. They were simply trying to bridge the gap until the time that guidelines could be established, until it could be ascertained whether the experiments could be safely undertaken, and until any necessary safeguards could be implemented. When they conducted their scheduled press conference on July 18, 1974, they expected little more than low-level headlines buried on the back pages of the morning newspapers, perhaps even forgotten by the afternoon. In that light, their shock at what did happen was understandable.

Field Day for the Press

The topic caught copywriters' imaginations in nearly every newspaper in the land. Out onto the page, in big, bold, black type, poured the most enthusiastic, flamboyant distortions imaginable. The headlines alone were enough to bring joy to the heart of any lover of science fiction:

Halt Genetic Experiments, Science Panel Urges
[Washington *Star-News*, July 18, 1974]

A New Fear: Building Vicious Germs
[Washington *Star-News*, July 23, 1974]

National Moratorium Urged on Hybrid Virus Research
[San Jose *Mercury*]

Hybrid Molecule Test Threat Seen
[Los Angeles *Times*]

Risk to Man Seen in Creating New Bacteria
[London *Times*]

Scientists Fear Release of Bacteria
[Oakland *Tribune*]

Genetic Scientists Seek Ban—World Health Peril Feared
[Philadelphia *Bulletin*]

Biologists Ask Ban on Two Tests—Chilling Theme
[Toledo *Blade*]

Scientists Renounce Gene Experiments
[Ft. Lauderdale *News*]

Bid to Ban Test Tube Super-Germ
[*The Observer*]

And if the headlines didn't satisfy, some of the prose was colorful and inaccurate enough to sell pornography in another setting, such as the following from *Chemical and Engineering News*:

Halt to Genetic Manipulation Urged

With images of an "Andromeda Strain" lurking in the background, a group of research scientists, in concert with the National Research Council [*sic*], has urged a moratorium on genetic manipulation of microorganisms. And general agreement by workers in the area that it is a good idea lends credence to the note of alarm. Paul Berg . . . called on workers currently doing research in genetic modification of bacteria and viruses to put off *all work* until a . . . task force can evaluate the hazards involved and recommend precautions for continued investigation . . . [italics added]

In retrospect, the scientists realized what had happened. Without providing enough background information on their debate, they had given the press a glimpse into the inner workings of the scientific community. And that peek, as small as it was, left a vivid impression of scientists, unaware of the consequences, battling secretly over an issue having immense implications for the public. Looking in from the outside, the press could legitimately ask whether the public shouldn't be the ultimate voice if scientists themselves could not decide if the research was dangerous.

If the research being discussed could indeed imperil the existence of life as we know it, should it be allowed to continue? In an issue of such importance, wasn't leaving the decision-making up to the practitioners, the very people who would benefit the most from the research—both professionally and personally—a bit like leaving war up to the generals?

Of course, the letter implied nothing of the sort. It supported the basic research. At the very most, it was an attempt to raise the level of awareness of scientists involved in what could be classified as potentially risky research, and scientists were faced with equally risky research in other fields every day. To scientists, recombination of molecules of DNA consists of a series of techniques. None of them are inherently evil, any more than a pistol or a knife lying untouched on a table can be called inherently evil. They are techniques, tools, pure and simple. Certain experiments that could be carried out with these tools seemed to carry the danger of possible risk; others did not. Their letter was an attempt to point out the dangers, to request that other researchers refrain from performing the experiments with the greatest potential for risk until an analysis of the true dangers could be made, and to set up the mechanism for analysis.

The dichotomy between the press and the scientists was broad, but not unbridgeable. It required only that the scientists educate the press, make available the information which would make their intentions obvious. But in their naivety, the scientists contented themselves with a press conference consisting of a simple question-and-answer session, implying that the moratorium was of unprecedented importance, and letting the letter speak for itself. And speak it did. With the publication of the Berg letter, the controversy entered its second phase. The first, characterized by small, intense, disorganized debate confined to the scientific community, was gone for good. The media had accomplished and what was known about the dangers follows. The controversy, like an egg, had hatched, and out into the world, squawking, fluttering, embarrassingly indecent, stepped the scientists' ugly duckling.

The rest of the world reacted well to the letter, although, before its publication, Zinder and Nathan had presented the idea of a moratorium to the participants at a meeting of the European Molecular Biology Organization (EMBO) meeting in Belgium and had received a mixed response. Some agreed with the letter's tenets; others fought them, claiming that if people would just stop talking about it, the problem would go away, and muttering angrily about this latest American conspiracy to corner the market on important new research. A moratorium would prevent the Europeans from trying to begin the research; it would not stop the Americans from continuing with it.

Reaction in England was slightly less favorable, partly because the British scientific publication *Nature*, which had published the letter in conjunction with its release in the United States, had cut the final paragraph in order to fit it all onto one page. The result was that English scientists read only the recommendations, without the humbling, cautionary caveats at the end. The arrogance of the abridged version offended them; it was not until Berg flew to England in September to defend the letter that either side became aware of *Nature*'s error.

The Aftermath: Preparing for Asilomar

The letter and press conference generated two distinct, but complementary, sets of actions within the scientific community. First, scientists all over the world dashed from their test tubes to their typewriters, unleashing an avalanche of passionate rhetoric on both sides of the issue. Editorials, articles, and letters poured onto the pages of otherwise dignified scientific journals—*Nature, Science, New England Journal of Medicine, JAMA, Genetics.* There were calls for the research to be placed under legal constraints, ethical constraints, no constraints at all. Some supported the conditions proposed in the Berg letter; others dismissed the signers as naive idealists who were abandoning promising avenues of research because of their emotional reactions to the pressures of hypothetical dangers.

The nucleus of the Asilomar organizing committee—

Berg, Roblin, and Baltimore—took a second tack, meeting immediately after the press conference to construct ground rules for the upcoming conference. Quickly they decided on the specific aspects of the problem that had to be discussed: the recombination of animal (eukaryotic) with bacterial (prokaryotic) cell DNA; the question of plasmids and how to control the transmission of genes; and the pathogenicity of, and dangers of working with, *E. coli.* They agreed that certain other questions had to be answered: Were the concerns justifiable? What information was indeed available? What could be done about protecting workers and public from dangers which did turn out to be real? Finally, they agreed to search for the best people to contribute to the discussion—those with the expertise to help in the assessment. They began to consider who should lead the various panels that would be set up to discuss each pertinent question.

By the time the organizing committee met again, on September 10, two new members had been added: Maxine Singer of the National Institutes of Health and Sydney Brenner, a British scientist who was recruited at least partly because his presence would dilute charges that Asilomar was an international conference in name only. It was this meeting, held at MIT, that dictated the final scope of the conference.

It was the consensus of the organizers that the one thing the burgeoning fields of molecular biology and genetics could not afford was a wide-open, haphazard congress of individuals, some of whom might pull the discussion far from the topic at hand, others of whom might dominate the conversation despite their comparative ignorance of the topic. Therefore, the committee agreed to convene a limited meeting, one restricted almost exclusively to those with specialized knowledge of genetic research. Strictly on a scientific basis, this decision limited their prospective participants to about 300 people around the world. The next criterion, just as crucial, required that scientific participants come from all parts of the world, to make the conference truly international; if any consensus could be reached, it had to be one with which every country could live. Without a broad consensus, and with the voluntary nature of the moratorium and the impossibility of actually overseeing the international research to ensure compliance, the conference would become nothing more than a charade, pablum to temporarily sate the appetites of the media and public, which were becoming ever more suspicious of the scientists' motives. Adequate representation implied agreement with whatever consensus might arise, and agreement signaled compliance. The process of selection was not going to be easy.

Determining who else should be invited raised other problems. Staff people from the funding agencies of both the conference and the research itself (NAS and the NIH) had to attend; representatives of those industries gearing up to give the research practical applications—pharmaceutical companies like Merck, La Petite from Italy, Searle from Great Britain, industrial giants like General Electric—all had to be included. And press participation, while it was finally deemed necessary (at first, to avoid a repeat of the fiasco surrounding the Berg letter, it was suggested that the press be barred from the meeting), was to be firmly restricted to sixteen places, filled only by those with a scientific background strong enough to ensure accurate, perceptive reporting.

In deciding upon the actual course that the conference would follow, the committee began by accepting several basic premises developed by Paul Berg. Berg felt strongly that the meeting must have limited, clearly defined goals. It should concern itself almost exclusively with the biohazards inherent in the basic research—research in the laboratory centering around the new recombinant technology. He was not interested in seeing the conference degenerate into a huge debate on the ethics of genetic engineering; nor was he willing to find it mired in discussion of such moot problems as biological warfare. To him and the rest of the organizing committee, the issue was research, pure and simple. Others could argue elsewhere about its ultimate implications.

In this vein, the committee decided to apoint three panels, assigning each the responsibility of evaluating the probability of risk in one aspect of the research. Richard Novick, who had signed the Berg letter, was placed in charge of the plasmid panel; its task was to examine the danger of transmitting genetic material through plasmids, the problems of antibiotic resistance, the natural history of plasmids and microorganisms, and the ways in which bacteria might pass their genetic material in the intestinal tracts of animals. Donald Brown, a genetic researcher at the Carnegie Institute in Baltimore, was chosen to lead the panel evaluating the risks of recombining eukaryotic (animal) DNA with prokaryotic (bacterial) DNA. Aaron Shatkin was to lead a panel examining the hazards of working with tumor viruses.

These three panels, which were to draw up reports for the conferees' consideration, composed the core of the conference. But, standing alone, they were scattered, diffuse. They played to the specific needs of the scientists, but did little to bring a social perspective to the conference. To fill this gap, the committee decided to ask Daniel Singer, Maxine's husband, to organize a small group to report on the full legal implications of the scientists' responsibilities. The creation of this final group met with less than enthusiastic approval from the entire committee; Berg and Maxine Singer opposed its formation partly because of the limitations of time, partly because it would be the only anomaly in what was, otherwise, a purely "scientific" meeting. But the legal aspects of risk were becoming more and more relevant to scientists as their medical colleagues began to run the gauntlet of malpractice suits. In addition, Congress, under the urgings of members as powerful as Senator Edward Kennedy, was threatening to look into the questions of legislative responsibility and the ethics of scientific research. After some debate, the committee agreed to the lawyers' session, placing it last during the three-day schedule. Both the timing and content of the session were to strike profound chords in the participants; without realizing it, the committee had provided for the conference a psychic wrench, one that could take the issues spelled out so neatly in the first days of the conference and bring their relationship with the rest of the world into sharp focus. The lawyers' session would develop into one of the crucial interpolations of the conference. Yet its very existence was an afterthought.

Before disbanding, the committee had one final task. It had decided upon the criteria for choosing participants: now it had to decide who would judge. Berg requested that each member of the committee send him a list of people that seemed eligible. He asked for similar lists from the leaders of the three panels. The lists could provide the nucleus of the conference; any small irregularities could then be worked out. But even this seemingly simple task raised problems. With the conference limited to about 150 participants (not including the press), at least half of the researchers with potential involvement in the research would have to be left out. And the requirement of international participation practically guaranteed that some places would have to be reserved for scientists with, at best, peripheral involvement, since few countries had actually begun research into the manipulation of genes. The Soviet Union, for example, was invited to send five representatives and did so. Yet none of the five had ever been involved in recombinant research; in fact, the Soviet Union, hindered by the bogus science of Boris Lysenko in the 1940s and 1950s, had not even begun research into the recombination of DNA molecules. The People's Republic of China, which was also sent an invitation, politely declined.

Limited, specialized participation caused other important points of view to be missing at the conference. The committee's agreement that only those involved in the research would be invited excluded groups questioning the validity of both the basic research then being undertaken and the applied research of the future. Groups like the Boston-based Science for the People (SESPA, from its original name, Scientists and Engineers for Social and Political Action, since the acronym was easier to pronounce) were not represented. Berg had actually tried to include SESPA—he had sent invitations to two of its leaders, Jonathan King of MIT and Jon Beckwith of Harvard, who had been part of a team that isolated a structural gene in bacteria in 1969 and had used the occasion of its discovery to publicize the growing exploitation of science by special interests. But King never received his invitation, and Beckwith was unable to attend; and when SESPA suggested someone to take their place, Berg rejected him as unqualified. The result of the selection system—and of the limited, specific subjects assigned to the three scientific panels—was that, of the 140 scientific participants, not

one was prepared either to propose prohibiting the
research completely or to discuss the greater questions of
ethics and control of the technology. These inherent limitations on the debate that was to come almost predeter-
mined its outcome. And that, after all, was what Paul Berg
wanted. Asilomar was to be a meeting focusing strictly
upon the problems of biohazards; the more amorphous
questions of ethics seemed to him to be as yet irrelevant
and insoluble.

And so the blueprint of the conference had been
sketched. Asilomar, that beautiful, forested site on the
coast of California, had been reserved for the end of February; the participants had been chosen, and their invitations were being sent; the panels had begun to organize
their reports and recommendations; and the problems surrounding the presence of the press had been at least partly
solved.

Meanwhile, elsewhere, the debate continued, growing
hotter by the day. In England, the Ashby Committee,
which had been set up in response to the Berg letter, was
holding hearings to "assess the benefits and hazards of
techniques which allow the experimental manipulation of
the genetic composition of microorganisms." After listen-
ing to testimony from both proponents and opponents of
the research, the Committee published its report in January 1975, barely a month before the Asilomar Conference
was to convene. The Ashby Committee was forced to
handle the issue with extreme care, for it encountered a
situation more explosive than that in the United States; in
March 1973, two uninvolved members of the public in
England had contracted smallpox and died after being infected by a laboratory technician who had been working
with the disease. The furor that followed was intense. Because the questions of physical and biological containment
of potentially hazardous biological organisms had already
become an issue of major importance, the Ashby Committee's report reflected a justifiable sensitivity. Although it
called for safeguards even more stringent than those proposed in the moratorium set forth by the Berg letter, it
stated that the hazards of genetic research seemed far less
serious than had been believed, that potential benefits were

"very great," and that, on balance, the "potential hazards need not cause public concern."

The curious convolution—dictating strong safeguards while declaring that the hazards were small enough to be practically ignored—was a solution not peculiar to the British. It has permeated the debate over recombinant DNA research ever since. As a solution it is a nonsolution, playing two sides against the middle, aligning the forces in favor of the research on both sides of the issue simultaneously. As a scientific stand, it is a failure; it weakens the claim of proponents that their position is based exclusively upon the "science" of the issue, and not on its emotional implications. As a politically motivated compromise it fares even worse, for it pulls the scientist from the pedestal of impartiality, leaving him vulnerable to any and every claim of opportunism. It expands the issue beyond pure science and permits opponents to ask two unanswerable questions: "If the research is safe, why will you agree to restrictions?" and "If you agree to restrictions, how can you claim that the research is safe?"

The answers to both these questions have much to do with the inadequacy of a purely scientific response to an issue as emotional as the controversy surrounding genetic manipulation. There existed an obvious need for pragmatic compromise. But scientists have seldom grown political wings and worn them comfortably. The Asilomar Conference, held, as it was, just as the controversy bloomed, was forced to at least try to accommodate several separate strands of the controversy which were bound, sooner or later, to join: public awareness of the power and freedom of scientific research, which had been growing since the debate surrounding the destructive and constructive uses of nuclear power, had finally reached a point where scientific responsibility was no longer assumed—the public was beginning to demand proof as a prerequisite before research could begin; the scientific response to the inevitable end of unfettered research was so obviously inadequate that it reinforced the public's awareness; and recombinant DNA research provided the ideal issue over which the greater questions could be fought. The problems of genetic manipulation, the specter of man having the power to change fundamentally both himself and the nature of the world around him, generated a love-hate response from the public. Recombination sat at the crossroads of the larger issue, emotional, ethical, and scientific TNT, waiting only for the right circumstances to set it off.

And so, by the time the Asilomar Conference arrived, the stage had been set for one of the most significant gatherings in scientific history, one whose implications far outweighed its actual work. To many of those attending, the intent of the conference was defined and restricted. To some of those who did not attend, and to others who observed, the issues were greater, and their subliminal, inevitable objections echoed throughout the actual proceedings. Like the revolutionary forms of DNA the scientists could create, the conference was becoming a hybrid, a chimera, displaying one face for everyone outside to see, while countenancing only in its more private moments and more serious conflicts its real personality and place in scientific history.

The Asilomer Conference, February 24–27, 1975

The scientists—the best of the world's molecular biologists—gathered in Asilomar as if they were under a state of siege. In a way they were. Their most precious blockade, freedom of inquiry, was slowly being toppled by the irresistible force of the growing awareness and demands of society. For only the second time in scientific history (the first occurred when scientists agreed not to disclose atomic discoveries to the Germans during World War II), scientists had placed restrictions on their own research, in the form of the Berg letter's voluntary moratorium. Now, after months had passed and planned research had been abandoned, they were finally convening to measure, as best they could, the hazards.

Some viewed the conference as a cleansing, a catharsis that would permit them to get back to their real work. The pause in restricted experiments would soon end; the discussion would lead to an informal statement endorsing the scientist's right to perform his research; hypothetical hazards would be exposed as hypothetical; the threat of outside intervention would be met by the clear, unified voice of the assembled group. The major question to some was: When can we start again?

Others were not so certain. True, the moratorium had not led to any revelation that the risks were real. But neither had it proved that they were illusory. In fact, besides bringing the momentum of the research to a jarring halt and protecting workers and public alike from the hypothetical dangers of ill-considered, sloppily performed research, the moratorium had done very little. To these scientists, the major question to be answered was altogether different: If, when confronted by the clearest evidence yet of science's immutable ties to society, scientists could not conscientiously regulate themselves, how long would society wait before it shouldered the burden itself through legislation?

The conference convened in Asilomar's small redwood chapel at 8:00 a.m. on Monday, February 24. It was opened by David Baltimore, a member of the organizing committee, who stressed the importance and goals of the conference. In one sense the meeting was the last resort for molecular biologists, for if they failed to come up with a consensus over the three and a half days of discussion, someone else certainly would. Baltimore's task was to make the intent of the conference clear, and he did so. He put the participants on notice that neither an apologia nor a whitewash would be enough to stem the future course of the blending of science and social policy. His statement extended the unprecedented tenor of the moratorium: precautions, never before instituted prior to the first occurrence of risk, would this time have to be imposed. For the first time, barriers to risk would be erected prophylactically, before tragedy struck, even if that tragedy could neither be accurately predicted nor measured. Perhaps politicians could wait for the first fatal crash before placing stop lights and cautionary signs at a blind curve in the road; scientists no longer had that luxury.

The statement was strong and unyielding. But it was merely the first salvo of the conference. Followed by a session devoted to disclosures of the state of the art of genetic manipulation, its impact was soon lost.

The first afternoon had been set aside for the researchers in the field to display their wares in a scientific show-and-tell. Presentations centered around what the research had accomplished and what was known about the dangers themselves. They also included a discussion of *E. coli* K 12, the universally used laboratory strain of *E. coli* that had been isolated from a single human gut some fifty years earlier. Although K12 is a weakened strain of the *E. coli* that commonly inhabits the intestine, researchers had confirmed that it could live long enough outside the laboratory environment to exchange genetic material with the hardier strains of *E. coli* which flourished naturally in the intestine. Confirmation of K12's capacity for survival came from British scientists, who showed a remarkable penchant for drinking concentrated beakers of the organism, then analyzing their stool for indications of its presence. Proof that K12 could survive outside the laboratory closed a crucial gap in the scenarios of the dangers of recombinant DNA research. It meant that any gene that had been recombined with a plasmid and placed in the *E. coli* host could, conceivably, be transmitted to bacteria living in the human gut, and from there to the humans themselves. And since *E. coli* K12 is the bacterium of choice in almost every genetic experiment, the danger of laboratory accidents and their consequences took on new importance.

It was not until Tuesday morning that the conference began to wrestle with the question of specific risks. At that point, the plasmid panel, under the leadership of Richard Novick, presented its conclusions in a thirty-five page, single-spaced memorandum which called for strict controls on the plasmid vectors used in recombinant experiments.

Novick was no stranger to the hazards of genetic research. In the early 1960s, in England, he had been working on a project to find a mutant strain of staphylococcus that was resistant to the antibiotic methicillin, a penicillin derivative. His search through staphylococcal mutants failed to unearth the desired strain, and Novick moved on to other research. It was only in 1974, when the problems of biohazards and recombinant research were surfacing, that he recognized the dangers of his old research and realized that, if he had found the mutant and had isolated it and bred it as he had planned, a laboratory accident might have led to natural staphylococcus picking up the resistance and destroying the usefulness of methicillin. The implications of this scenario were worse than they seem on the surface; in the late 1950s and early 1960s, a worldwide epidemic of penicillin-resistant staphylococcus attacked hospital patients in every corner of the globe. The epidemic killed scores of patients and weakened thousands more until hospitals under siege were forced to fire doctors, nurses, and other medical personnel who carried the bacteria. "When we couldn't kick the staph out of the doctor," said one hospital chief, "we were forced to kick the doctor off the staff." It was only with the discovery of methicillin and related drugs to which staphylococcus was still sensitive that the epidemic was slowed. Even now, although staph is kept well under control in most hospitals, staphylococcal infections are a major factor in the startlingly high rate of hospital-caused infections that exists throughout the world.

Novick had formed his committee from the personnel working on a project to standardize the nomenclature used in plasmid research. The group, composed of Novick, Stanley Falkow, Stanley Cohen, Roy Curtiss III, and Roy Clowes, had already established a working relationship before their new project began, and they functioned well together. They met first on November 7–10, 1974, to hammer out a position paper on experiments involving the use of plasmids as vectors for bacterial host systems like *E. coli*. In a series of eighteen-hour sessions, they devised the simple principle that has characterized the guidelines on recombinant research ever since: a graded series of risks demands a graded set of precautions to match them.

The panel's final document recommended that possible experiments be ranked in six classes, according to their expected degree of hazard. Class I implied no hazard, where experiments could continue freely, whereas Class VI contained hazards so severe that no physical containment procedures would ensure the reduction of biohazards to an acceptable level. Such experiments were to be banned entirely.

The plasmid panel's recommendations precipitated the first crisis of the conference. The previous day's activities had consisted mainly of researchers trotting out their toys for display, and harmony was easily maintained. But the plasmid panel's report posed a threat. Denunciations and attacks came from all sides, and one English scientist— Ephraim Anderson—engaged in an angry exchange with Novick during the discussion of the report:

ANDERSON (*reciting*): For our purposes, pathogenicity and virulence are defined similarly as "the capacity to cause disease." This must rank as the greatest oversimplification of all time. (He then questions the qualifications of the entire panel and the value of a 35-page, single-spaced report.)

NOVICK (*apologizing*): This is, after all, a rather terse document.

ANDERSON: You could have fooled me.

NOVICK (*paraphrased*): Alterations can be made. After all, this is a working document. We only worked on it for six days.

ANDERSON: And why couldn't you do it in six days? After all, the Lord created the world in only seven.[1]

Joshua Lederberg, a Nobel laureate, attacked the document for its seeming precision and inflexibility, arguing that it would be easy for a legislature to translate it directly into law, a tragic outcome considering the profound ignorance of whether the hazards really existed at all. And Sydney Brenner, the British scientist serving on the organizing committee, repudiated the implication of authority that pervaded the document, saying that the real issue was how to avoid presenting any risk to scientists, laboratory workers, and public alike. He continued:

. . . If people think they are going to get a license from this meeting, a notice they can put up on their door, if they are just pretending there is a hazard and are going along with it just so they can get tenure and be elected to the National Academy and other things that scientists are interested in doing, then the Conference will utterly have failed.[2]

With the discussion following the plasmid panel's presentation, many scientists began to demonstrate a remarkable professional instinct for survival; the conference

[1] Michael Rogers, "The Pandora's Box Congress," *Rolling Stone*, June 19, 1975.

[2] Nicholas Wade, "Genetics: Conference Sets Strict Controls to Replace Moratorium," *Science*, March 14, 1975.

threatened to disintegrate into a divisive squabble among factions, with each trying to salvage its own particular brand of research. At that moment, Sydney Brenner saved the day with a call for discussions into the problems of biological containment. According to him, if the products of the research could be successfully kept within the laboratory, without the threat of accident and subsequent escape by potentially harmful hybrid bacteria and plasmids, the basic question of biohazards would be resolved.

Two types of containment are possible: physical containment and biological containment. But physical containment is a matter for engineers, and its success still depends upon the conscientiousness of the researchers using it, since if they are blatantly sloppy or careless, no physical barrier can prevent the inevitable accident. Biological containment is a different story. It is based on the fact that if the vector (plasmid or bacteriophage) and the bacterial host (for example, *E. coli*) cannot live without the special conditions of the laboratory, neither a breach of physical containment nor negligence on the part of the researcher will cause a potentially dangerous escape of hybrid organisms. Brenner therefore called for meetings on Tuesday afternoon of those interested in creating "safe bugs," hosts and vectors so weak and so dependent upon the laboratory that their vulnerability compensates for the inadequacies of physical containment and the hypothetical biohazards that may be caused by escaping organisms.

With Brenner's sessions came an atmosphere of peace. The conference seemed to have found a way around the prickliest threat facing it: that research would be cut off indefinitely while methods of safeguarding public and ecosystem were created. The search for safe bugs was an acceptable, wholly possible enterprise. It was also fine, elegant science. With the passing of this threat, the conference again settled down to the task at hand.

Through Tuesday and Wednesday, the conferees examined each of the three panel reports, discussing their recommendations. Neither of the remaining two reports matched the competence or the thoroughness of the Novick report. Shatkin's group, which was to examine experiments that involved using animal tumor viruses, such as SV40, as vectors for the hybrid DNA, produced little more than a

collection of the letters that members of the panel had sent
to each other. That panel's conclusion—that experiments
using animal viruses as vectors could proceed safely under
the existing National Cancer Institute guidelines for handling oncogenic viruses—was vigorously disowned in a minority report submitted by one of its members, Andrew
Lewis.

Lewis had, in 1969, isolated an accidental, laboratory-
created, hybrid cancer virus, the product of natural recombination between SV40 and adenovirus 2, a member of a
common, usually harmless family of viruses found in human beings. When Lewis reported on his find at a meeting
at Cold Spring Harbor in 1969, he did so with certain reservations. He recognized that the hybrid virus presented
possible hazards of unknown consequences: nobody had
any idea of its virulence, its ability to infect, or even if it
could be harmful to other organisms. He also realized that
a new, heretofore unknown virus that had been created in
a laboratory would generate tremendous interest among
the researchers performing cancer research. And Lewis, as
the discoverer, was responsible for deciding to whom he
should distribute his new virus.

The meeting at Cold Spring Harbor did little to ease his
burden. After his talk, in which he expressed reservations
about distributing such a potentially dangerous substance
to anyone who did not prove willing and able to use the
proper safeguards, he was accosted by James Watson,
the Nobel laureate, who accused him of trying to hoard the
precious virus for his own scientific advancement. Lewis
was taken aback; Watson, as one of the brightest stars in
the world of molecular biology, was the last person he had
expected to react emotionally to what seemed to him to be
a considered evaluation of potential risks. And Lewis became uneasier as his task progressed. He did finally decide
to distribute his virus, but only to those laboratories that
agreed to comply voluntarily with certain safety precautions set out in a "Memorandum of Understanding and
Agreement" that he had drafted. Later, after distributing
the virus, he discovered that several laboratories that had
agreed to the requirements of the memorandum had failed
to fulfill them, both in establishing safeguards within their

own walls and in restricting distribution to those who complied with the memorandum.

Lewis presented his minority report late Tuesday afternoon and received a far more favorable reception than the majority report. By now, most of the scientists participating had resigned themselves to the need for some form of self-regulation; even though Lewis' experience with precisely the kind of voluntary regulation being proposed had been a failure, his recommendations—that some aspects of the use of viruses as vectors be subject to more stringent safeguards than those already in effect—struck a responsive chord. The conference, using Lewis' report as a catalyst, jumped all over the majority report. Late in the day, the panel was confronted with a question from one scientist that drove to the heart of the matter:

"Since you have proposed that current reagents such as SV40 may be used, and since you have proposed an immediate end to the moratorium, what experiments do you plan to do when you go home Monday morning?"

The question was a perfect counter to the contention by James Watson and others that, since the moratorium had uncovered no real hazards in the research, it had served its purpose and should be terminated. The panel members were trapped; not one had the nerve to advocate actually using animal viruses as vectors before new precautions had been arranged, either by the National Cancer Institute or by some other quasi-legal body. And so they confessed that they themselves did not actually plan to perform the research.

"Why?" the panel was then asked. "Are you worried about it?"

The members admitted that they were. And the contradiction in their position became clear. Their report had nothing to do with the advisability of using animal viruses as vectors; rather, its conclusions were based on the scientists' philosophical belief in freedom of inquiry. The problem they had left unresolved was that, given the potential for danger arising from the research, it might be unrealistic to expect each and every researcher to act responsibly, on his own, in controlling its use. Their answer to Brenner's question made it clear that even they doubted it.

The session on animal tumor viruses ended soon

thereafter. The panel pulled itself together for the first time since it had been formed and met, day and night, drafting a new statement, which it submitted to the organizing committee late on Wednesday night. The revised statement, in essence, supported the position taken by Andrew Lewis' original minority report.

The final report, submitted by the panel led by Donald Brown, consumed most of Wednesday morning. The class of experiments evaluated by the eukaryote panel had not been embargoed by the Berg letter; the signatories had asked only that work proceed with caution. Yet it covered a group of experiments—the so-called "shotgun experiments"—that had become as great an issue as those of tumor viruses and plasmids. In shotgun experiments, the entire complement of DNA of an organism is exposed to restriction enzymes and chopped into recombinable lengths. The resulting fragments are then inserted into bacteria and grown. These experiments are by far the easiest type of recombinant experiments to perform, since they require no process of selection of the "correct" gene to be studied. All genes are cloned, each within its particular fragment of DNA. The researcher then simply isolates the one that he wishes to work with after the recombination has been successful. The danger is simply that, by taking the blind, shotgun approach, the researcher might be cloning, in addition to the desired genes, genes carrying the code for a previously repressed tumor virus. With its fragment liberated from the restraining influence of the rest of the DNA, the recombined tumor virus might then grow, escape, and infect man. And further recombination might induce human genes to pick up the activated tumor gene, replicating it with its own, normal genetic material and making it a part of the genetic complement of the human race.

Because of these dangers, Brown's panel placed shotgun experimentation at the top of its scale of hazardous experiments. In both scholarship and caution, its work fell somewhere between that performed by the plasmid panel and that of the tumor-virus group. After a session marked by a distinct dampening of the passion of the previous afternoon, the recommendations of the third panel were sent on to the organizing committee.

With the conclusion of the session on the recombination of eukaryote with prokaryote came the end of the technical aspects of the conference's agenda. Brenner's groups, still working on ways to construct safe bugs, continued their work throughout Wednesday afternoon. And the organizing committee, which was to draft the summary statement of the conference and had begun its work late Tuesday night, was also kept in almost constant sessions. The only meeting that had yet to be convened was the lawyers' discussion of the legal aspects of the controversy. Most of the participants looked forward to it as a slightly inconsequential diversion, a welcome interruption to almost three solid days of hard, emotional science. They could not have been more wrong.

The lawyer's session was the cement that annealed the random bricks of the conference. As Daniel Singer, its organizer, commented, it "moved things from being a little diffuse into focus. It responded, oddly enough, to an incredible sense of unease that permeated the people who were there." Unease indeed. The limited scope of the previous sessions, which had kept discussion confined to issues of fact, had allowed the conferees to arrive at a basic agreement on the three panel reports. It had also somehow accommodated most of the conflicting points of view, permitting each to settle more or less comfortably into the framework of the conference. What it had not achieved was a legitimate, practical perspective that made sense of the myriad issues involved. In writing its summary statement, the organizing committee was having difficulty pinpointing the direction of the conference, primarily because no concrete direction had yet appeared. And solving problems of fact, no matter how satisfactorily, could not be enough for a group which knew instinctively that more was at stake. By raising the precise "peripheral" issues the scientists had been consciously and unconsciously trying to ignore all along—the issues of ethics and legal liability, of the researcher's and science's evolving responsibility to society—the lawyers' session became a guiding light for many conferees and ultimately set the conference's tone.

The first lawyer to speak, Daniel Singer, was by far the least offensive. Playing on the theme of responsibility and its meaning for the researcher, he covered the previous

three days in fine generalities, discussing them from the
perspective of the ethical obligations of the scientific investigator. Singer's talk caused little pain. After all, he was
merely reiterating what many of them had been saying all
along. Their primary responsibility was to do no harm.
Further, all agreed that the ethical issues were indeed complex, justifying, perhaps, a conference at a later date.
When Singer sat down, few of the participants had any
inkling that things were about to become very rough.

The second speaker, Roger Dworkin, was a professor of
international law. Dworkin had been assigned the task of
describing to the participants what the law on scientific
research is, and how it might affect their work. His main
premise broke the thread of noble purpose and condescen-
sion held by some of the scientists present, those who were
willing only to agree to conditions that provided an external semblance of regulation, a cosmetic set of guidelines.
Scientists were being permitted to attempt self-regulation
because of the respect that they had garnered over the
years, Dworkin said. But that freedom was a gift from the
public; if their work demonstrated even the appearance of
indulgence, legislatures, in the name of the public, would
undertake to exercise legal right to control anything that
might harm the public. Simply put, "It is the right of the
public to act through the legislature to make erroneous decisions."

The implications were clear. The issue had come to a
conflict of rights, and the right of the public to be safe
from harmful research superseded the scientific community's right to freedom of inquiry. The power to perceive
when the right of safety was threatened lay, ultimately,
with the public—through its elected representatives—
whether or not its perceptions were accurate. If the Asilomar Conference—which was, after all, a conference of last
resort, convened to decide the fate of the voluntary moratorium—applied cosmetic, unrealistic guidelines under
which the research might continue—or even if the public
simply perceived the guidelines as cosmetic—the scientists
would have nobody to blame but themselves, and no
recourse when legislatures began passing laws designed to
control their work effectively. Their only hope was to

assess the hazards realistically and to take a firm stand on the real need for regulation.

The third lawyer to speak, Alexander Capron, also brought scant comfort. His talk—on the specific institutions which hold power over the continuation of recombinant research—pointed out that not only did the legislature have the right to regulate research, but that it was the federal government, whose purse strings are held by the legislative bodies, which functioned as the researchers' major source of funding. The public pressure upon these funding institutions to respond to a danger of unknown magnitude could very well cut off the flow of money, rendering even legislation irrelevant, at least for the university scientists. In addition, areas of government as far afield as the Labor Department held power over the continuation of research through existing legislation; for example, the Occupational Safety and Health Act (OSHA) requires that the work place of laboratory technicians be free of hazard. Its standards are set by the Secretary of Labor.

With the lawyers' session, many participants began to realize the truth: they had not been called together merely to put forth a magnanimous document conceding certain irrelevancies to a critical few, while reserving the true right of research for themselves. This was a serious business. To an extent, it involved the survival of the way science had been carried out over the centuries, an issue far greater than the simple matter of evaluating the risks of a certain methodology. If the scientists did not feel it was their moral or ethical responsibility to act, demonstrating a real concern and commitment to safety would, at the very least, be one of the more astute political concessions they could offer.

While the rest of the conference left after the lawyers' session to attend a small beer party, the organizing committee met to complete the statement they would present to the final meeting the next morning. They worked and wrote through the night. When morning came, they had their statement. But they were concerned about the possibility that the conference might still be recalcitrant, might be unwilling to agree to a set of controls that was actually stricter and more precise than those set out in the original

Berg letter. And so, in a telling move, the organizers decided to try to prevent the rest of the conference from voting on their provisional statement.

As the final meeting got under way on Thursday morning, it became clear that the mood of the conference had
shifted dramatically from the day before. Although arguments arose over specifics, and while the provisional statement was debated point by point, the participants were
ready to recognize the need for decisive action. Indeed, as
the meeting progressed, the major point of consensus came
on the belief that these, the first clearly defined recommendations, should be drawn strictly in the face of scientific
ignorance. If new information showed them to be
needlessly confining, they could then be relaxed. A vote of
the entire conference on the statement, which the committee had dreaded, was in the end forced by the conferees.
But the committee's fears proved groundless; it lost only
one vote, when the conference decided to specify, in addition to the gradations of low-, moderate-, and high-risk experiments, a group of experiments that could not be
performed under any circumstances.

At noon the final vote was taken. The provisional statement was adopted, and the organizing committee was authorized to mold it into a final draft that conformed to the
conference's wishes. In that provisional statement, stuck to
its end like some vestigial tail, was the symbol of the true
value of the meeting, as well as an indication of its frus-
trations. The document concluded with six unanswered
questions, each one crucial to a real evaluation of the
risks, each eloquent testimony to how little the scientists
really knew:

1. Are eukaryotic genes or viruses expressed in prokaryotic
   hosts and, if so, can they modify the bio-hazard potential of
   these cells?
2. Can free DNA molecules infect animals or plants?
3. Can prokaryote-eukaryote recombinant DNA molecules,
   either free or encapsulated as phage particles, infect animal
   or plant cells and be expressed there?
4. Can mammalian cells in culture be genetically transformed
   by free homologous or heterologous DNA?
5. Can hybrid animal virus DNA or virus-plasmid hybrids
   cause tumors in animals?

6. Can methods be developed to monitor effectively the escape and dissemination of cloning vehicles?

These six questions posed the final paradox and irony, the ultimate Catch 22 for the meeting: Without answers to these questions, how could any reasonable guidelines be instituted? And without reasonable guidelines, how could the research into these questions be undertaken?

The Summary Statement of the Asilomar Conference

The provisional statement underwent slow revision before being published on June 6, 1975, in *Science*, *Nature*, and *Proceedings of the National Academy of Sciences*. If the conference was to be judged on the basis of a comparison between the organizers' original goals and the contents of the statement, there could be little doubt as to its success. But whether the goals set by the organizers were appropriate was a far different question, one that has attracted passionate debate ever since.

The statement began by describing the principles that had guided the conferees in their deliberations. First, and most important, came the question of ending the seven-month moratorium on certain types of experiments, if "the scientific work could be undertaken with minimal risks to workers in laboratories, to the public at large, and to the animal and plant species sharing our ecosystems. . . . The participants at the Conference agreed that most of the work on construction of recombinant DNA molecules should proceed, provided that appropriate safeguards, principally biological and physical barriers adequate to contain the newly created organisms, are employed."

Because the moratorium was being called off, the conference had devised principles for dealing with potential risks. First, containment must become a part of the experimental design of each recombinant experiment; second, the degree (and effectiveness) of containment "should match, as closely as possible, the estimated risk." Unfortunately, the provisional statement never explained how the estimated

risk could be accurately assessed without the answers to
the questions it posed at its conclusion.

Having described its principles and the methods used to
determine them, the committee proceeded to outline four
levels of physical containment—minimal risk, low risk,
moderate risk, and high risk.

Minimal-risk containment, intended for experiments
where biohazards could be accurately assessed as minimal,
required nothing more than the standard procedures used
in regular clinical mirobiological work: "no eating, drinking, or smoking in the laboratory, wearing laboratory coats
in the work area, the use of cotton-plugged pipettes or
preferably mechanical pipetting [instead of pipetting by
mouth], and prompt disinfection of contaminated materials."

Low-risk containment, for experiments "which generate
novel biotypes but where available information indicates
that the recombinant DNA cannot alter appreciably the
ecological behavior of the recipient species, increase significantly its pathogenicity, or prevent effective treatment
of any resulting infections," added little to the containment of minimal-risk experiments, other than a sign on the
laboratory door demanding that the uninitiated keep out,
and the use of safety cabinets in experiments producing
aerosols.

Moderate- and high-risk containment conditions were
confined to facilities far more escapeproof. They also required the use of vectors and hosts that, for moderate risk,
were "appreciably impaired" and, for high risk, were
"rigorously tested [organisms] whose growth can be confined to the laboratory."

Moderate-risk containment, "intended for experiments in
which there is a probability of generating an agent with a
significant potential for pathogenicity or ecological disruption," also required physical facilities with biological safety
cabinets (laminar flow hoods), vacuum lines protected by
filters, and negative air pressure within the laboratory itself,
to ensure that particles would not escape on air currents.

High-risk containment, "intended for experiments in which the potential for ecological disruption or pathogenicity of the modified organisms could be severe and thereby pose a serious biohazard to laboratory personnel or public," added isolation by air locks, treatment systems to decontaminate everything—air, liquids, and solids, animate and inanimate—leaving the laboratory, and chemical showers for the personnel.

In theory, the containment package permitted most experiments. In practice, it maintained the moratorium on experiments with high-risk components, since no rigorously tested and approved safe bugs yet existed. The "appreciably impaired" bugs that were acceptable for moderate-risk experiments were also guaranteed to provoke debate, since the preferred host for recombination, *E. coli* K12, was significantly weaker than its free-living relatives, yet had shown the crucial capacity to survive long enough to exchange its genetic material if it did escape.

After having described the possible levels of containment, the organizers defined the types of experiments that fell under each level. In doing so, they used the same three categories first defined in the Berg letter and then characterized by the three panels at the conference: animal viruses, eukaryotes and prokaryotes, and bacteriophages and bacterial plasmids. The experiments were classified by means of their assumed degree of risk; those most directly affecting man, such as experiments conferring new types of antibiotic resistance or those that recombined segments of DNA from warm-blooded animals, were afforded the highest levels of containment. But even a detailed set of recommendations like these begged the real questions surrounding risk while the quality and actual danger of the risk remained unknown. And it was left up to those who demanded more stringent controls to suggest that experiments be divided into those which contained obvious benefits and those whose risks obviously outweighed their potential for good. Then, beneficial experiments, regardless of the risk, could be performed under the highest containment levels possible—affording real protection to the public—while all other experiments could be banned. This system of evaluation would have been easier to put into

practice, since accurate assessments of the precise level of
risk for each experiment would no longer be necessary. In
addition, it functioned on the logical principle that *any*
potential hazard, whether real or imagined, should be coun-
tered with the best containment available. But this train of
thought ran directly counter to the conference's own principle of matching the degree of containment to the estimated risk. And although trying to match containment to
risk is like trying to compare oranges to apples, the policy,
in practice, permitted a flexibility and freedom that the organizers felt was crucial at this preliminary stage of evaluation.

A fourth type of experiment was to be deferred because
the dangers inherent in it were so serious that, even with
the proposed host-vector systems, the theoretical hazards
remained unmanageably great. This type included recombining genes from the most highly pathogenic organisms
known to man—which produce the botulinum and diphtheria toxins of the world—recombining DNA which contained toxin-producing genes, and performing experiments
on such a large scale that containment was practically impossible (this included using more than 10 liters of culture, a level that had little effect upon the research
scientists but was certain to impede the work of scientists
in industrial and pharmaceutical laboratories, who would
require far greater amounts of culture to meet demand
and generate profit if marketable recombinants were
created).

The statement concluded with a summary of the conference's findings on safe bugs, a reiteration of the principal
investigator's responsibility to train his staff in safety precautions, and a short outline of the avenues of research yet
to be undertaken that would lessen the chances of the accidental escape of hybrids. The six questions—which had
looked so formidable at the end of the provisional statement just a few months before—were buried, incorporated
into the body of the text.

In this way, the conclusions of the Asilomar Conference
were presented to the rest of the world. The organizers
had succeeded in capturing most of what the conference
had wanted to modify in that final Thursday meeting. But
certain small discrepancies did surface in the final state-

ment. The conference had agreed to define three types of containment (low-, moderate-, and high-risk); the organizers had added a fourth (minimal-risk) which downgraded the risk and containment of certain experiments, even though the differences between minimal- and low-risk containments were negligible. Also, the conference had implied that moderate-risk containment should include host-vector systems with no capacity to multiply outside the laboratory, rather than the "appreciably impaired" capacity discussed in the statement. Finally, the designations of risk in each experiment were to have been so strictly drawn that any revision of containment practices would result in relaxation of the procedures; instead, the final statement viewed the evaluation of the experiments as "interim assignments which will need to be revised *upward or downward* in the light of future experience" (italics added).

Since each of these three discrepancies weakened the conference's original recommendations in some way, they were troublesome. The last one, especially, was bound to cause problems, for there are only two conceivable ways that containment procedures would need to be revised upward: either experimentation would have had to show that the experiments were more dangerous than had originally been assumed and were literally bursting the seams of a facility with weak containment procedures, or containment at a lower level would have already had to have been breached and an escape would have had to occur. Despite these problems, the basic tone of the conference—the urgency implicit in its being called in the first place, the influence of the lawyers' session, the knowledge that cosmetic guidelines would not be enough to prevent controls from being imposed from the outside—had been echoed. In reaching at least the limited goals originally envisioned for the meeting, the conference had succeeded beyond what had, at first, seemed possible.

But was Asilomar a success or a failure? Its specific recommendations were superseded within a year by more detailed guidelines from external agencies. The level and quality of its discussions were soon bypassed by new knowledge, by protagonists with greater control of the sub-

ject and its implications. The Asilomar that will remain
history, then, is not the conference presented in the tight,
turgid summary statement of the organizing committee.
For Asilomar was a conference of people, a gathering
where, for the first time, scientists debated as best they
could the implications of their new ability to fundamentally alter life. It was both ironic and touching that their
awakening should take place on the shores of the Pacific,
the place where, millions of years ago, life began.

The Press

"Before they showed up, some reporters didn't know a
plasmid from a Pacific seagull," said one of the participants at Asilomar. But by the time they left, they had
asked enough questions, shoved enough microphones into
enough faces, and listened to the cacophony of enough
scientific jargon to qualify as experts on the controversy.

The press role at Asilomar was unique. Partly because it
had been burned by the reporting of the Berg letter, and
partly because of an innate discomfort with the motives
and perceptions of the media, the organizing committee
for the Asilomar Conference had treated the question of
press participation seriously from the beginning. Unlike
some of the participants at the conference itself, the members of the committee were wide awake to the potential
and the danger of the influence of public awareness. How
the press perceived the conference and its conclusions was
almost as important as the conclusions themselves, for if
the public could be shown that scientists were acting responsibly in regulating problems emanating from their
work, the specter of external control could be restrained,
at least temporarily. And the way to gain the respect of
the reporters was not to hoodwink them—given the structure of the conference, with 140 headstrong scientists
ready to speak at any time on any subject which seemed
relevant, it would have been impossible anyway—but to
find a way to educate them, to keep them from filing their
stories prematurely, to practically make them as much a
part of the conference as were the scientific participants. If

the press could be made to appreciate, in detail, exactly what such a conference involved, perhaps its reactions could be trusted.

To accomplish this, the committee decided upon a historic series of rules for the press. First, nobody could file a story until the conference was over. Reporters would not be dashing for the phone every time someone made an inflammatory statement during a discussion. It would also give each member of the press time to educate himself fully on the real issues being debated. The methodology of recombination could become clear; the potential hazards could be perceived in the correct perspective; and the potential benefits could be seen as perhaps something more than figments snatched from the fantastic world of the scientific imagination. Second, no reporter coming to the conference could arrive after it had begun or leave before it had ended. This reinforced the embargo on premature stories. It also gave each reporter an enforced period of grace, an argument to counter the deadline mania that infects the editorial staffs of so many publications. Third, the press corps would be restricted to sixteen members, to be chosen by the organizing committee. In fact, the committee could probably not have enforced this rule if it had been challenged—one reporter did threaten to take his case to court if he were refused entry—and four uninvited reporters ultimately attended the conference. But the combination of rules restrained many publications from sending reporters. And those that did—*Science*, the New York *Times*, the Washington *Post*, the Los Angeles *Times*, the San Francisco *Chronicle*, and *Rolling Stone*, among others—were many of the very publications that the organizers preferred having represented.

Despite the care taken by the organizing committee, the initial colloquy between the scientists and the journalists was less than auspicious. During David Baltimore's opening remarks, a small forest of cassette tape recorders sprang up on the edge of the stage. Some scientists complained that they had been told the meetings could be held off the record, and it was only by a show of hands that the participants approved the journalists' right to use recording equipment. Thereafter, members of the press showed re-

markable prudence, keeping silent during the formal
sessions and swamping scientists with questions only during off hours or more formal debriefing sessions. The one
break in the protocol arranged for the press, by an Associated Press stringer who shot off a quick story to the San
Francisco *Examiner* on Tuesday, caused a minor flurry,
but the other journalists present held to their promises and
the conference proceeded without further incident.

The presence of the press brought to the conference an
aura of importance that it might not otherwise have had.
Because the press contingent was so large—the ratio of
press people to scientists was about one to eight—a sense
of history pervaded the conference. Simply because it was
there, the press discouraged backroom deals and blunted
some of the grandstanding that might otherwise have infected the conference. By the end of the conference, the
journalists' proximity was so natural that Stanley Cohen—who had shielded his face from photographers at the
beginning of the conference like some convicted hood—responded to a Senate subcommittee's charge that public advocates had had no voice at the conference by pointing to
the press as the participating arm of the public. His contention was, of course, absurd; the members of the press
were limited to their traditional roles as observers. They
asked no questions at the formal meetings, made no statements, never participated in the voting or the creation of
the summary statement. They became, in fact, faceless
symbionts for the scientists, feeding upon their knowledge
even as they provided welcome relief and perspective during the intensity of the four-a-day meetings. But the very
fact that Cohen made the suggestion indicates how well
the press played its role. It, too, benefited from the rules
imposed by the organizing committee. With no daily dead-
lines to meet, the journalists had time to sort out their im-
pressions and construct more complete, more accurate
pictures of what had occurred; the pieces they wrote were
models of restraint and reason, demonstrating a full appreciation of the dilemmas facing the participants and of the
depth and scope of the problems at hand. As reporting,
some of the pieces published were gems of the journalistic
art: Michael Rogers' piece in *Rolling Stone,* for example,

and Stuart Auerbach's long dissertation in the Washington *Post* were both fascinating and well reasoned.

The handling of the press was one of the minor victories of the organizing committee. After the debacle of the press treatment of the Berg letter, the scientists had learned to minimize the irrelevancies and irresponsibilities of ignorance—a problem that had been caused not by the press but by the scientists' initial inexperience in communicating with the outside world. With the widening involvement of the public in scientific matters, it had become indispensable for them to learn to translate their difficult, often incomprehensible language into words that at least a journalist could understand; he could then be the one to clarify the complexities of science for his readers.

Asilomar thus became one of the first instances where a difficult scientific problem became accessible to the layman. And so a fourth element began to fuel the controversy surrounding the manipulation of genes: to a public growing increasingly aware of the fragility of its own existence, to a body of professionals adapting poorly to the requirements of a relationship undergoing rapid, fundamental change, and to an issue as emotional as it was real was added the means for communication that gave anybody with the desire to learn an opportunity to join in the debate, if not on the techniques themselves, at least on the issues that the techniques created. Informed public scrutiny of science had become a real possibility. The Salem witch trials were not going to be reenacted over the controversy of genetic research; the danger of the majority, driven by ignorance, fear, and prejudice, tyrannizing the minority had become remote. With the vehicles of communication available, it was now up to both sides to use them.

Reaction to Asilomar

There is little doubt that, by some standards, the Asilomar Conference was a complete success. It fulfilled the goals of its organizers and their sponsors; it solved the immediate problem of the moratorium; it generated the momentum that would later be manifested in the specific, defined guidelines drawn up by the major funding agency for recombinant research, the National Institutes of Health. But success is as relative a concept as deciding whether a graying sky should be called partly sunny or partly cloudy. And while the chorus of appreciation for the work of Asilomar was loud and long, other voices, on both sides of the issue, disparaged its results.

By a dozen criteria, the conference had fallen short. The participants had failed to come up with any meaningful way to actually measure the risks of the research, if indeed such measurement was at all possible. The summary statement, a document of remarkable honesty, showed that little new knowledge had surfaced since the onset of the moratorium in July 1974. In trying to maintain a distinction between the science of the issue and the ethical implications, the participants had left themselves open to the lawyers' argument that such a distinction was irrelevant, immaterial, and unrealistic, for the question of ethics was closely tied to any issue that threatened the public good. And the influence of the lawyers' session in pulling the conference to a final consensus did not change the fact that nothing in the summary statement responded directly to the lawyers' most significant recommendation: that the scientists, whether they liked it or not, had to subordinate their own research to the welfare of the public whenever the two clashed.

Those public interest groups studying the issue noted that, of the 140 scientists present, not one was an epidemiologist, a representative of a union (whose members, working in laboratories, would be directly affected by the decisions), or a member of the group of governmental watchdogs working under the aegis of the OSHA. As one member of Science for the People (SESPA) observed, "It would be asking too much to expect a group of people getting together to protect themselves [from] the public to have representatives from consumer groups [present]." Asilomar looked suspiciously like a conference composed exclusively of those whose best interests would be served by cosmetic guidelines, guidelines which did the least to inhibit their work. By setting out to limit the conference to purely scientific questions—and by devising a set of regulations which effectively dealt only with the technical aspects of the problem—the participants had implied agreement on a flood of other problems, none of which had really been settled at all. The problem of pure research, for example, was handled within the confines of the conference by the unspoken consensus that it was affected solely by the issue of public safety; if pure research does not harm the public, it is automatically permissible, since practicing beneficial science is the birthright of the scientist. Although recombinant research is pure, the reasoning went, it contains theoretical hazards, and therefore must be considered in a different light. But groups like Science for the People were ready to challenge the belief that pure research even exists. Research is not performed in a vacuum, they argued. It is a part and parcel of the social, political, and economic setting of the society in which it flourishes. The only possible way to ensure the performance of "pure" research would be to make unlimited funds available to anyone who wanted the money for any research project with a suitable, practical protocol. In this society, with governmental agencies providing the main sources of funding, and with the evaluation boards of those agencies selecting the types of research that seem most promising, there is no possible way to avoid social judgments. Pure research is thus nothing more than an ideal, a concept not suitable for the real world. Anybody who professes to think differently is either naive or dishonest.

Other great issues—the potential dangers of application of the research, the hazards of recombinant work being performed on a large scale in private industry or in military establishments, the questions of genetic surgery and microbiological warfare—were also ignored by the conference. Its singleminded goal of determining the specific dangers of the basic research, to the conscious omission of issues which would inevitably infect the controversy, left it open to extensive criticism.

Opponents of the research did not have a monopoly on the criticism. Scientists arguing for the right to continue their work could find as many holes in the restrictions recommended by the conference as others could find in its omissions. Without a single clue as to how to measure the risks, the conference had placed restrictions on the research. Without a single statement as to what the dangers are—these "novel," "unknown," "potential" dangers that would arise from recombination—the moratorium was effectively maintained and, in certain instances, expanded. A hardnosed analysis of the events that would have to occur to make the theoretical risks real was never undertaken. As a result, the entire conference was predicated on misinformation and a lack of experimentation, and resulted in recommendations developed not from true scientific assessment but from fear of the political intrusion of Congress if affirmative action was not taken. It was not only a witch hunt from the press and public that the scientists had to fear; the potential for oppression existed within the ranks of the scientists themselves, among those who would rather bow to external pressure than stand up for hard, factual evidence. And the evidence is clear: there is no proof that the hypothetical hazards discussed at Asilomar could become reality. Several of the biological processes that would be necessary to turn even the most virulent recombinant organisms into threats to man are not even known to exist; even if they did, the odds on their occurring together, in the correct order, are so small as to be a mathematical improbability. Furthermore, a discovery in itself is a piece of knowledge, a tool, and does not have the capacity to be inherently evil. Society's penchant for putting neutral scientific advances to destructive uses has nothing to do with the scientists themselves, whose task it is merely to discover, and not to judge. Finally, it is not enough to simply say that some questions—otherwise scientifically valid—are so frought with danger that answers to them should not even be sought. The only way to determine whether the hazards do indeed exist is to perform the research, in a central location, with the most advanced physical and biological containment and the finest microbiological techniques available. Then, if the hazards are shown to exist, the dangerous material can be destroyed on the spot; if they are shown to be purely theoretical, recombinant research can brush off the stigma and the work can continue.

Both sets of arguments contained a superficial validity

which made them impossible to discard out of hand. And emotions within the scientific community were running so high that the most vocal of the protagonists had begun to resort to personal attacks on their opponents to make their points. Despite Asilomar, it became obvious that the debate would continue, for the lack of hard evidence permitted anybody to set the risk-to-benefit ratio at whatever level best suited his position. The issue itself was growing beyond the confines of science and infecting the larger arena of politics; and whether the scientists liked it or not, the emphasis was slowly shifting from the problems of containing the basic research to the larger questions of ethics.

The nonsolution provided by the Asilomar Conference struck the ostensible unity of the molecular biologists like a boulder splashing into a placid pool. Even as its importance as a milestone in the history of science's relationship with society grew, its specific recommendations became as temporary as the moratorium the year before. A strict analysis of the actual risks had yet to be achieved, and the problems of providing containment with enough strength to protect against danger and enough flexibility to permit the continuation of the research had found no firm solution. The controversy had obviously grown too large and too complex to be up to informal, *ad hoc* meetings of the experts. And so, with the summary statement of the Asilomar Conference in hand, the real overlords of the scientific community, the holders of the purse strings, set in motion the events that would lead to a definitive set of regulations. In November 1974, the National Institutes of Health created the Recombinant DNA Molecule Program Advisory Committee to take up where Asilomar was bound to leave off, to function in a small, analytical group, and to arrive at detailed, precise guidelines that the NIH and other agencies could use as criteria for distributing grant money. This time nobody was going to have the luxury of voluntary compliance. The teeth behind the guidelines would be small—suspension of funding both for the principal investigator and for the institution supporting his research—but they would be sharp. And the existence of even this tiny threat implied that greater penalties could be imposed if it proved insufficient. The controversy surrounding research into recombinant DNA molecules was moving to a new phase, from the hands of the researchers themselves to a codified set of regulations administered by a group that was sympathetic, but not involved in the research. The public and its spokesmen had begun to flex their muscles.

THREE:
The Benefits

Although it was the technique of recombining molecules of DNA that triggered the debate over the place of science in society, recombination is but one aspect of a profound and general revolution that has swept through biology during the past thirty years. The advances that have occurred have been possible because scientists have discovered ways to do strictly defined experiments with the simplest organisms available—the viruses, bacteriophages, and bacteria —by using genetic and biochemical tools simultaneously. It has been this newfound capacity to do delicate experiments in relatively easily understood organisms that has led to the discoveries of the nature of genetic material, the way genetic messages are transmitted, the mapping of nucleotide sequences, and the details of protein synthesis.

But it was the technique of recombination—and not the myriad other discoveries—that finally touched off the controversy. By making possible the direct manipulation of nucleotide sequences—or genes—in DNA, recombination took man beyond the primitive ways he had previously directed genetic change and gave him the opportunity to play God. Before recombination, scientists trying to take advantage of genetic changes in plants or animals had to wait for the natural process of random mutation and biological recombination to occur before they could choose among newly mutated characteristics and increase the proportions of those characteristics they wanted to maintain. Now, chemical recombination may enable scientists to deliberately transfer desirable sequences of nucleotides from one species to another, bypassing the natural order of mutation and selection completely.

But discussing the capacity of recombination to permit scientists to bypass the natural order of selection is stretching the initial debates too far. Recombination is simply a series of techniques, and nothing more. Although it is potentially one of the most powerful tools available to biologists, it exists not as a philosophy or principle, but as a new, possibly indispensable way of increasing our understanding of how we grew from a single cell to an almost infinitely faceted, billion-celled organism, with hands where hands should be, a pancreas that does what a pancreas is supposed to do, arms, liver, brain, all in the right places and performing the right tasks. Like a hammer for a carpenter, recombination is there, available, ready to be used.

The crucial question, then, in its simplest form, is: Will the use of this new technique do us more harm than good?

But measuring the ratio of risks to benefits is not as simple as drawing a line down the middle of a sheet of paper, listing benefits under one heading, risks under another, and seeing whether they balance. It is, in fact, ridiculous to try to evaluate the relative worth of a potential cure for cancer against the relative danger of a worldwide epidemic of hyperinsulinism. Furthermore, since the technique is so new, and since all discussion centers vaguely around the future, the odds against solid evidence providing answers are overwhelming. And with the inherent unpredictability of science and basic research (for science that knows where it is going is no longer basic science at all, but the application of science and the development of technology), scientific protagonists in the debate are left with precious little in the way of directly relevant facts. They must base their views on a mountain of extraneous fragments of information that can be made to fit any position and the almost incomprehensible mixture of intuition and experience that goes with each person's own philosophical bias. Meanwhile, those outside of science who have involved themselves in the controversy have neither experience nor the same balance of information to depend upon; they have to replace these ingredients with social intuitions, long-term views of the situation, a sense of the inevitable conflict between the welfare of society and the freedom of scientific inquiry, and a clear understanding of the larger implications of the debate. It has been said that 20 percent of all scientists can predict accurately about 50 percent of the time, while the rest are almost never right. Comparable ratios have not yet been computed for the public.

To complicate the issue still further, it is difficult, if not impossible, to measure the value of the techniques of recombination without analyzing the potential of the larger field of genetic research as well. It is not enough to define recombination as a way of creating novel organisms, for, although it can do precisely that, that is not its primary value in research. Recombination allows the scientist to isolate a specific segment of DNA, plug it into a bacterium like *E. coli*, and grow it until he has enough material to work with. The technique is designed to make a small DNA factory of *E. coli*, to make it practical to take perhaps 1/1,000 of the DNA complement of an organism and use it experimentally. Once the DNA is isolated, grown, and purified in large enough quantities, the real experiments into the secrets of genetics begin.

Evaluating the risk-to-benefit ratio of recombinant DNA research, then, is not simply a matter of examining the techniques themselves. Without a clear understanding of what the techniques make possible, without a recognition that basic research affects not only the laboratory work itself but also the results of that work, without the vision to see both beneficial and harmful potentials of the basic research, hypothetical as they may be, it is impossible to evaluate accurately the true implications of the techniques themselves. And just as this holds true for the issue of basic science, it also applies to the larger question of whether even research as potentially risky as recombination provides strong enough grounds for our interfering in the right of science to search freely for knowledge. Ultimately, the issue of recombination cannot be isolated from its philosophical and ethical consequences and examined only on scientific grounds; for the debate surrounding it will be played time and time again, as other aspects of science and technology reach levels of sophistication where they, too, can touch directly the bare nerves of the public good. Unless standards are set at the beginning for evaluating these inevitable controversies, some segment of society, whether consumer or professional group, is going to be stepped on.

There are obvious benefits to be had from the use of recombinant DNA technology. Some appear in the field of basic research; many are in the process of being realized right now. Although it has existed for just a few years, directed recombination has already been a major factor in the advancement of fundamental knowledge. In addition to helping scientists understand the structure and function of genes, recombination has provided information about the structure of those plasmids which cause antibiotic resistance in bacteria, about how bacteria propagate, how they evolve, and how their genes are regulated. While in the past our inability to isolate specific genetic regions has limited our understanding of how the genes work in the more complex cells, recombination has provided information about how genes are organized into chromosomes and how gene expression is controlled. It is information like this which will make it possible for scientists to discover how defects in the structure of such genes can alter their function and cause genetic disease.

Other benefits, in the fields of applied research, are mainly speculative, grounded more in scientific prediction than in fact. Still others, even farther in the future, have barely made the transition from science fiction to quasi-legitimate scientific speculation. While speculation is a scientific habit, and a necessary one, speculation in biological research has its own special problems. Unlike the physical sciences, biology has very few "first-principle theories"—theories which are recognized to be basic and universally true for all aspects of the science—upon which to base predictions. The immutable physical laws of nature—from gravity to thermodynamics—are well known and reinforced daily; chemical laws of action and reaction are also as solid as rock. But biology's first-principle theories can be counted upon the thumb and forefinger of one hand. First, there is the theory of evolution. Then there is the recognition that nucleic acids are the basic genetic material, present in all forms of life. And that is just about all. Biology is the science in which the words "never" and "always" never appear, for almost every rule ever devised except those of evolution and DNA seems to have its exceptions.

Because of the basic genetic value of DNA, an understanding of its workings is crucial to an understanding of the entire controversy. No real comprehension of either benefits or risks is possible without a knowledge of how DNA determines our genetic characteristics, of how, step by step, a small, linear series of bases along the length of a chromosome can determine the color of eyes, the texture of hair, or the size of ears.

How DNA Works

Everything in the body is ultimately the product of the action of enzymes. Enzymes, which are actually functional proteins, have the capacity to catalyze a series of chemical reactions which finally produce literally every characteristic that has a genetic foundation. The pigment melanin, for example, makes skin tones dark: the more melanin that is produced, the darker the skin tone. But if that chain of enzyme reactions leading to the formation of melanin is cut—if even one of the enzymes necessary to melanin's production is not formed—the organism involved will lack melanin, lack skin pigmentation, and develop albinic characteristics.

It is DNA's function to carry the code which produces both the enzymes and the structural proteins that are the prerequisites of life. The way in which it does so is extraordinarily simple and elegant, a testimony to the penchant of evolution for finding the most direct, most efficient solutions available when it is faced with complex problems.

In its most simplified diagrammatic form, DNA looks

structurally like a ladder. Its two backbones are composed of deoxyribose (a sugar) and phosphates, its rungs of the four bases, adenine, guanine, thymine, and cytosine. If DNA were divided into two separate, single strands, each would be constructed in precisely the same fashion; the bases are joined to the sugars, while the base-sugar units (the nucleotides) are joined by the phosphates. In double-stranded DNA, the two identically formed strands are bonded together according to the rules of complementary bases: adenine pairs only with thymine, and guanine pairs only with cytosine. The final, simplified structure looks like this:

D = deoxyribose
P = phosphate
T = thymine
A = adenine
C = cytosine
G = guanine

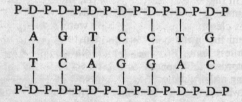

This structure continues for the entire length of the chromosome, a distance which may include hundreds of thousands of base-pairs, with alterations only in the sequence of the bases. The actual DNA molecule is helical in shape, spiraling from end to end and making a complete revolution around its axis every ten nucleotides.

How DNA replicates is obvious from its structure, as Crick and Watson learned. How it passes on complex genetic information is not so clear, although it is evident that the way the bases follow each other in sequence contains the information, rather than the steady, unchanging chain of sugars and phosphates.

For the sake of clarity, let us construct our own, imaginary segment of DNA, and follow it through the process that leads to the formation of protein and, ultimately, life:

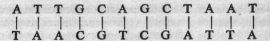

These twelve nucleotides would not be enough to form even one of the tiniest proteins; a typical code for a small protein might consist of at least 100 nucleotides, while the smallest viruses contain over 5,000. A typical single-cell bacterium might contain about 10 million nucleotides; and the genetic code contained in the chromosomes of a mammalian cell may have as many as 5 billion nucleotides, or about a million times as many as the tiny viruses.

In the early 1960s, Francis Crick and his colleagues delved into the incredibly large, seemingly incomprehensible patterns of these nucleotides. They found that DNA did not transmit its information in long, unbroken lines but in small groups of three nucleotides each, called codons. The method was a little like Morse code, with three places to fill and four symbols to choose from. Thus, even in the longest strands of DNA, there exist only a limited number of possible combinations—4^3 or 64 codons. It was clear that 64 combinations were hardly sufficient to contain the amount of data needed to create even *E. coli*, much less man; but if these combinations acted as individual words in a sentence, and if the sentences could each be as long as would be required to create one protein, the system would work. Proteins are composed of long, linear chains of amino acids. And each three-nucleotide codon was found to carry the genetic message for either an amino acid or for portions of the stop-start signal that controls its release.

Our own hypothetical segment of DNA thus contains four codons, specifying four amino acids: ATT, GCA, GCT, and AAT (the complementary nature of the second strand makes it unnecessary for us to refer to it, for, like replicating DNA, we can reconstruct it at will).

While DNA exists in the chromosomes, proteins are manufactured by small protein factories, the ribosomes, which float around in the main body of the cell. To send its genetic message from the chromosome to the ribosome, the cell uses the simple vehicle of ribonucleic acid—RNA, DNA's cousin.

RNA comes in three forms, each with its own specific function: messenger RNA (mRNA), which accepts the message from the DNA of the chromosome and delivers it to the ribosomes; transfer RNA (tRNA), which transfers the information brought by mRNA to the correct amino acid; and ribosomal RNA (rRNA), which, with protein, makes up the ribosome itself. All three forms have the same basic composition. Like DNA, they are linear structures, but they are composed of only a single strand. Each strand is composed of a sugar-phosphate backbone and four bases; but RNA's sugar is ribose (instead of DNA's deoxyribose), and the thymine base of DNA is replaced by a new base, uracil. A typical RNA molecule looks like this:

R = ribose
P = phosphate
U = uracil
A = adenine
C = cytosine
G = guanine

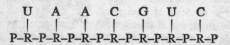

The first step in the translation of genetic information into actual proteins occurs when DNA transfers its message to mRNA. Since the message for each individual protein is contained in a linear segment of the long DNA molecule, each message is supplied with start and stop mechanisms which tell the mRNA how much of the molecule to read off. At the point where the message begins, the two strands of DNA begin to separate into two distinct strands, as the bonds between the complementary bases disintegrate. The separation continues to the end of the message, at which point the molecule again becomes a double-stranded unit. The separation of the two strands leaves a part of the DNA molecule available to link up with new complementary bases. And that is exactly what happens. An enzyme, RNA polymerase, fishes free-floating bases out of the surrounding cytoplasm and brings each to its respective partner. For our molecule, this means that the DNA chain of ATTGCAGCTAAT will receive its complementary chain of UAACGUCGAUUA. Because the required sugars and phosphates are already attached to the free-floating bases, all that is necessary is for the RNA polymerase to anneal the gaps in the sugar-phosphate chain, finishing the new RNA molecule (Figs. 7,8). When it is complete, the newly formed mRNA peels off of the DNA molecule like a label from its backing, and the DNA molecule reforms its original bonds. The new

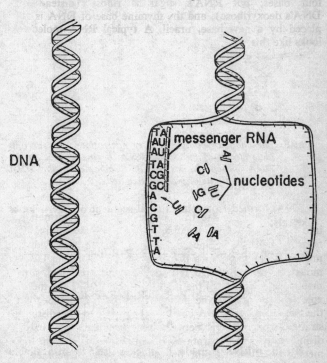

FIG. 7. *Translating the genetic message. A protein "reads" the start code of a gene, causing the DNA to divide and making each of its strands accessible to complementary nucleotides.*

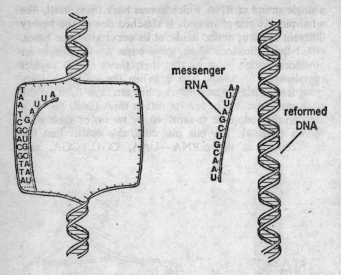

Fig. 8. *The chain of complementary nucleotides, now a complete strand of messenger RNA, begins to peel off the DNA. When the separation is completed, the DNA reforms into its original double helix.*

mRNA, formed from our original segment of DNA, looks like this:

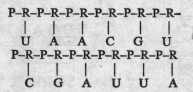

After the mRNA is made, it attaches itself to an available ribosome.

Meanwhile, individual strands of transfer RNA have been floating freely in the cytoplasm, picking up the specific amino acids for which each has been coded. tRNA is a single strand of RNA which curves back upon itself, like a hairpin. At one of its ends is attached one of the twenty different kinds of amino acids; at its curve are three bases, called the anticodon. These three bases are known as an anticodon simply because, together, they form a perfect complement to a single codon. When the mRNA has arrived from the chromosome and has attached itself to the ribosome, the tRNAs begin to attach themselves, one at a time, to the codons of the mRNA by means of their anticodons (Fig. 9). For our molecule, this means that the four codons of the mRNA—UAA, CGU, CGA, and

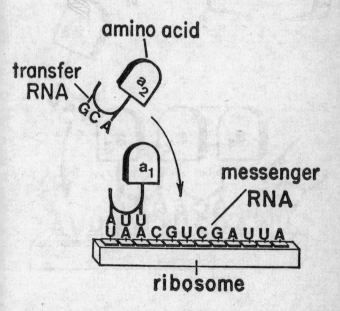

FIG. 9. *Messenger RNA attaches itself to a ribosome. Individual units of transfer RNA, each with its three-nucleotide anti-codon, read the codons of the messenger RNA. Their complementary bases attach.*

UUA—attract tRNAs carrying the anticodons AUU, GCA, GCU, and AAU. When two tRNAs have attached to the mRNA, the amino acids attached to their ends form bonds by means of enzymes. Then, as the first tRNA (AUU) detaches from its amino acid and drops off the mRNA, the third tRNA (GCU) attaches itself to the mRNA and bonds its amino acid to the first two. The process continues until the entire message of the mRNA has been read off by the tRNA (Fig. 10).

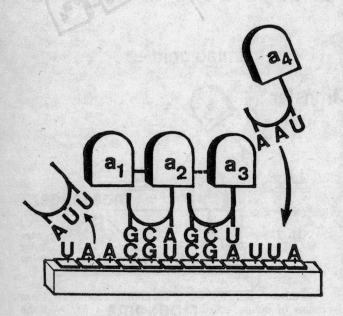

FIG. 10. *The amino acids attached to each transfer RNA begin to form a linear connection. The process of attraction of transfer RNA, connection of amino acids, and discard of used transfer RNA continues until the entire message has been read.*

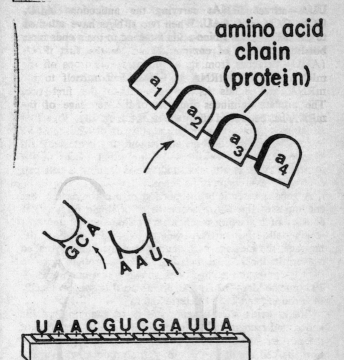

FIG. 11. *The process is complete. The original genetic message contained in the sequence of nucleotides of DNA has been transformed into a finished chain of amino acids—a protein.*

In this way proteins, which are nothing more than linear chains of amino acids, are formed (Fig. 11). And the small segment of DNA with which we began has caused the formation of a chain of four amino acids, each one's position and character determined by the codons of the original DNA and mRNA and the anticodons of the tRNA. In similar ways, much longer chains of amino acids form the enzymes that catalyze the chemical reactions between substances to form and reform molecules, break down and rebuild structures, and tie functionally complementary materials together until a cell is formed, then an organ, skin, teeth, glands, until the organism is complete and functioning.

Understanding the Genetic Structure

The relatively recent discovery of the way in which genetic information is transformed from message to actual structure or function is only a small part of what must be learned before practical genetic engineering is available on a wide scale. One of the most important pieces of information that is still not understood is how a cell can control the expression of its genes.

A gene is essentially the portion of a chromosome that encompasses the entire sequence of a single message. It consists of a starting mechanism (called the "promoter") which tells the mRNA where to begin reading the message, the message itself, and a stop mechanism (called a "terminator"). The message for the protein tyrosine tRNA, for example, has been shown to be composed of 199 nucleotides—52 in the promoter, 126 which specify the product, and 21 in the terminator.

But knowing what a gene consists of, mapping specific genes, and recognizing their products out of the context of the normal functioning of a cell tells us very little about how the cell regulates the use of the product, when it decides to manufacture it, and why. And questions like these are not merely on the periphery of necessary genetic knowledge; they are at its hub.

The reason that the cell's regulatory mechanisms are so important is fairly simple. Each cell of an organism contains its entire complement of DNA; thus, each cell contains the potential to perform any and all of the reactions necessary to induce life and function. Yet every cell in a multicellular organism like man suppresses all but the smallest fraction of its genetic messages. In this way, some cells become muscle cells, some blood cells, others nerve cells; some become parts of the torso, some parts of the head, others parts of the extremities. Despite their carrying identical complements of DNA, the cells' differentiation is so great that, both in function and in structure, they are practically unrecognizable as clones of each other.

Although we have learned at least part of what regulates certain reactions in certain cells, we do not yet have the slightest understanding of how these regulatory mechanisms are controlled to the incredibly delicate degree necessary for making us the complex organisms that we are. We have practically no understanding of why the code for manufacturing insulin is expressed in the pancreas—and only in the pancreas—of every normal human being, and why it is suppressed everywhere else. We have yet to locate the places along the chromosome that supply the more intricate regulatory messages; nor do we even know in what form those messages are transmitted.

The problem is further complicated by the fact that, in addition to the linear component that it might be expected to have, the regulatory mechanism must also have a recognition of time and space. Working out the components of the simple linear reaction necessary for translating genetic messages into proteins thus does little more than scratch the surface, for the regulatory system controls not only the precise moments that each reaction must function, but also the amount of material it must produce, and the extent to which its material reacts with the other cellular substances before it is again suppressed. Looking for the mechanism of an on-off switch is easy; finding out what does the flicking, and when, why, and how, is an entirely different story. The complexity of the regulatory mechanism can be seen in the amount of DNA needed to control it. It has been estimated that only about 5 percent of the DNA in higher organisms, including man, is used to actually produce proteins; the rest is devoted to regulation.

Much of the speculation surrounding the future of recombination and genetic research in general depends upon our success in discovering how the cell controls its regulatory mechanisms. For that reason, many of the experiments now being conducted are tied to achieving that goal. Seen in that light, it is obvious that recombination, with its capacity to manufacture quantities of purified DNA, is indispensable if research is to continue at full speed. Otherwise, it is inconceivable that the applied goals of genetic research, such as identifying the genetic bases for birth defects and cancer, can soon be reached.

Mapping the DNA Molecule

Before scientists began to delve into the specific components of the regulatory mechanism, it was crucial that a method be developed that permitted a quick reading of the DNA message—a way of mapping the sequence of the nucleotides.

In 1976, two such methods were devised, the first by Dr. Frederick Sanger of Cambridge University in England, the second by Drs. Gilbert and Maxim at Harvard.

Sanger realized that, while most DNA is double-stranded and divides into two single strands only when it replicates, a small group of viruses are composed of single-stranded, circular molecules of DNA. They become double-stranded only after they invade a host cell, when they rob the cell of some of its nucleotides to complete their own DNA chains. These single-stranded virus DNAs seemed perfect for a method Sanger had devised for determining the sequence of nucleotides, since they were already in a state where each nucleotide along the chain was ready to bond with its complementary nucleotide.

Sanger therefore took several of these single-stranded DNAs and set them aside. Then he took other samples of double-stranded DNA from the same virus and cut them into segments by means of restriction enzymes. The resulting segments were separated by size through the process of electrophoresis, which forces particles in a gel to migrate in response to an electrical field. After he turned the electrical current off, Sanger extracted tiny segments of DNA whose sequences he already knew to prepare the single-stranded DNAs for analysis.

When Sanger combined the single-stranded DNA with the chopped-up, double-stranded segments, the single strands attracted the segments and bonded with them at the appropriate complementary sites. After the segments had annealed, Sanger added to the test tube a loose mixture of all four nucleotides, with at least one tagged with radioactive phosphorus, and the enzyme DNA polymerase. With the DNA polymerase acting as a catalyst, the appropriate nucleotides began attaching themselves to the small primer, reconstructing the complete DNA chain. After a short time, Sanger chemically halted the growth process. Since the reconstruction had continued at different rates for each of the DNA strands in the test tube, Sanger now had a number of partially reconstructed strands, with at least several stopping at each nucleotide of the strand he was mapping.

Sanger then split his sample into four subsamples. Into each of them he poured a different mixture of nucleotides (with a different nucleotide missing from each mixture) and DNA polymerase; in this way, he covered all the possible combinations needed to continue the DNA's growth. The strands whose growth he had halted began growing again. But this time the strands in each of the four samples stopped their reconstruction whenever they reached a point at which the missing nucleotide was needed to continue the chain, for DNA reconstructs itself sequentially, adding one nucleotide after another to the existing chain. Because the original strands had had their growth chemically halted at every possible point along the segment of DNA he was studying, each subsample would now contain strands that had stopped growing at every point where the missing nucleotide was required.

Again Sanger used electrophoresis to separate the partially completed molecules according to their size. After they had migrated, and after he had taken an x ray of the results, he could simply read from the film the points at which each of the strands had stopped growing. If, for example, a mixture of guanine, thymine, and cytosine had been added to one sample, the DNA would have stopped recombining at each point where an adenine was needed. After performing this process with each of the four samples, Sanger could map the segment of the DNA that he was studying.

Dr. Sanger's original methods of sequencing were slow and cumbersome: it took nine researchers two years to uncover the 5,375 nucleotides of the first small virus that was mapped. Already, those original techniques have been modified so that two researchers have been able to map a similar virus within one year.

The Fantastic Quirk of Phi X-174

The first virus mapped by Dr. Sanger's methods was Phi X-174, a tiny bacteriophage that infects *E. coli*. Identifying the entire nucleotide complement of Phi X-174 permitted scientists to determine the precise roles played by its parts. Phi X-174 manufactures nine proteins, four of which make up the coat that surrounds its minuscule squiggle of DNA. The other five proteins are functional proteins which concern themselves with replication and assembly of the virus as it proliferates inside *E. coli*.

In determining the messages which coded each of the nine proteins, the scientists made an amazing discovery. Contrary to the popular dictum that had been handed down since the early 1940s (which maintained that each gene produced one, and only one, protein), they discovered that there was an overlap in the linear messages of Phi X-174's DNA. Instead of each gene containing a promoter, message, and terminator, several of Phi X-174's genes actually began one message during the middle of the preceding message:

codons of message one

....C A T T C G C C T A C G G G C T T T G C....

codons of message two

The result is a little like a crossword puzzle, where the correct solution must fit in several directions at once. Since proteins are developed only from very specific codes, and since the majority of nucleotide sentences composed at random would be meaningless, it is an extraordinary occurrence when an organism as tiny as Phi X-174 develops a message as contingent upon the luck of the draw as this one is. The odds against its happening are tremendous.

Since Phi X-174, several other viruses, including SV40, have been shown to have similar overlaps. It is assumed that such overlaps are a function of the size of the virus, that this particular way of developing its message was more efficient for the organism, more suitable to its size and environment. Most scientists doubt that the phenomenon carries over into higher organisms, where the problem is usually that there seems to be excessive and redundant genetic material, rather than too little.

The discovery of Phi X-174's overlap destroyed the consensus that the one-gene/one-enzyme theory was a first-principle biological law. It is a shining example of the dangers of too much anticipation and speculation in a science as prone to exceptions as biology. In the instance of Phi X-174, something that seemed well beyond the capacity of genes has been demonstrated to occur; it could just as easily happen that something that seems well within the potential of the future may actually be impossible.

First Attempts at Solving the Regulatory Mechanisms

The secrets of the cell's regulatory mechanisms are only slowly coming to light, mostly through painstaking laboratory research and model-making. But certain advances have clarified some of the simpler ways that these mechanisms can work.

In the early 1960s, François Jacob and Jacques Monod of France devised a model that explained one way that regulation could occur. Their success was one of those shattering breakthroughs that remains obscure and unrecognized outside the scientific community. But within it is viewed as one of the most brilliant scientific models ever devised.

Jacob and Monod took the sequence of DNA that codes for the enzyme beta-galactosidase, studying the entire gene—from promoter to terminator—to determine exactly what it was that suppressed the enzyme and what triggered its production. Normally, if the gene were just a free piece of DNA, the RNA polymerase enzyme that catalyzes the construction of mRNA would recognize the start signal, attach itself to the gene, and start reading the sequence of nucleotides, bringing free bases from the environment to form a complementary chain. However, in the bacterial system in which they were interested (which just happened to be *E. coli*), Jacob and Monod discovered a gene existing in front of the gene for beta-galactosidase which somehow acted as a regulator of its function: when that gene was triggered, beta-galactosidase was not produced; when it was suppressed, the ribosomes began to manufacture the enzyme. They discovered that the regulator produced a protein called, in their system, a repressor protein. The repressor protein had a lovely affinity for the tiny area between the attachment of RNA polymerase and the start mechanism for the beta-galactosidase gene. It simply sat there, acting as a roadblock and preventing the RNA polymerase from reaching the start mechanism and reading the sequence of the gene.

But if the repressor protein were simply allowed to remain on the chromosome, and if the cell did not have a mechanism for its removal, the cell would forever be without beta-galactosidase. In most cases with cells from higher organisms, that would be perfectly fine, for most genes in those cells remain permanently repressed. *E. coli*, however, needs beta-galactosidase. It therefore has its own mechanism for removing the repressor. Jacob and Monod found that when the simple sugar lactose entered the cell, it would interact with the repressor protein to pull it off, permitting the cell to manufacture beta-galactosidase.

The beauty and simplicity of this system became even clearer when Jacob and Monod realized that beta-galactosidase acts on lactose. The cell could not afford to waste its own energy making the enzyme unless lactose was present, so it devised a way to use the presence of lactose to release the enzyme. And lactose, acting automatically to pull off the repressor protein, becomes the active agent of its own destruction. When the lactose has been destroyed, the repressor protein can once again attach itself to the beta-galactosidase gene and suppress the manufacture of the enzyme.

The Jacob-Monod model is a wonderful example of the symmetry of function and logic that is a part of the workings of genes. The manufacture and suppression of beta-galactosidase is a perfect closed system upon which the cell spends only the smallest amounts of its energy. In it, scientists have begun to understand the intricate mechanism behind regulation.

Chemical and Biochemical Synthesis of DNA

After the first discovery of how the start-stop signal worked for at least one gene, scientists began to look toward artificial synthesis of DNA as a way of uncovering important information about its structure and function. Two methods of synthesis have been devised. As alternatives to recombination, both have significant drawbacks which prevent them from being used on a large scale in the production of large amounts of usable genetic material.

The first method's problems are best symbolized by the twenty-four scientists who took nine years (a total of 216 man-years) to chemically synthesize a single gene. Although the process has been streamlined and is now far more efficient, it is still extraordinarily time-consuming and tedious.

The process was first devised by H. Gobind Khorana, who had by then transferred from Wisconsin to MIT. To perform a chemical synthesis, Khorana decided to follow the previously mapped nucleotide sequence of tyrosine transfer RNA, the gene with 199 nucleotides. Slowly, painfully, he built blocks of nucleotides in sequence and annealed them with DNA ligase. His technique was similar to that developed by Frederick Sanger for mapping nucleotide sequences; he would bring all the necessary components of each reaction together, chemically block all reactions except the one he wanted to take place (with substances that could be readily removed from the chemical broth), and then activate the newly formed group so that he could locate it and transfer it to the next reaction. When the gene was complete—with promoter, message, and terminator—he introduced it into *E. coli* and found that it functioned as it should in its new environment.

Khorana's method required hundreds of steps for each chemical reaction, with a poor ratio of success to failure for each step. In addition, the entire process had to be kept in perfect synchronization; if the rhythm of the experiment deviated only slightly from what was necessary, the partially completed gene would refuse to grow.

Although the process of chemical synthesis requires that

the sequence of nucleotides for the gene being produced be known, it also means that specific alterations can be introduced into the structure, achieving many of the same results achieved by recombination.

Biochemical synthesis is far less tedious than chemical synthesis. But its potential is also more limited. Devised by Thomas Maniatis in Cold Spring Harbor, biochemical synthesis involves beginning the process with mRNA (the natural product of DNA in the synthesis of proteins) and working backward to form DNA by means of enzymes. This process of reverse transcription can quickly synthesize segments of DNA; its major drawback comes from its ability to synthesize only the part of the message read by mRNA. The promoter and terminator, two crucial areas on every gene, are not read off in the transcription process, and therefore do not appear in the reverse synthesis. Although the lack of the start-stop signals probably prevents Maniatis' synthesized DNA from being expressed in host bacteria, his approach does yield useful DNA samples. When compared to identical genes functioning in the cell, they will enhance our understanding of the structure and function of the eukaryotic gene.

Eukaryote vs. Prokaryote: The Final Hurdle for Speculation

Although the increased knowledge of the past decade has brightened considerably the landscape of applied research and technology, prediction within the realm of probability is still restricted to those advances which revolve around single-celled organisms. The reason is that recent research has been performed almost exclusively on tiny microorganisms, which have only 1/1,000 the amount of genetic information that resides in mammals such as man. Isolating a gene that is only one of several thousand is difficult enough; isolating that same gene when it is only one of several million is infinitely more difficult.

Additional problems arise from the different characteristics of bacterial and animal cells. Bacterial cells, or prokaryotes, are far smaller than eukaryotes (animal cells) and have far fewer structures and systems. One structure that prokaryotes lack—the structure that is actually used to define the difference between the two types of cells—is a true nucleus.

The lack of a nucleus means that the chromosomes of prokaryotes float around in the cytoplasm of the cell in the immediate vicinity of the ribosomes. Therefore, protein synthesis, the object of the transcription of DNA's messages, is a direct affair, proceeding simply from DNA to mRNA to tRNA to an amino acid chain. In eukaryotes, the mRNA, constructed in the nucleus, has to somehow pass through the nuclear membrane to contact the ribosomes. Scientists still have no idea how the cell undertakes its transportation.

In addition, prokaryotes seem more efficient and far more precise than eukaryotes. For example, animal cells can manufacture only ten of the amino acids necessary for life, drawing the rest out of their environment; prokaryotes like *E. coli* can synthesize them all, plus all the water-soluble vitamins they need. Furthermore, 90 percent of the RNA manufactured in the nucleus of a eukaryote never even reaches the cytoplasm to interact with the ribosomes; instead, it turns itself over, breaks down in the nucleus, to be synthesized again. What is all the extra RNA being synthesized for? Scientists do not understand its role. But from what they can tell, the process is simply a precious waste of the energy of the cell. For eukaryotes, with their tremendous excess of genetic material, it may simply be a case of the careless rich who have not yet been brought to heel by that ultimate master of regulation, evolution. Prokaryotes, the poor, miserly cousins, permit no such waste; since they must synthesize everything themselves and can rely on no outside sources of nutrients, they cannot afford the massive indifference to excess that seems to be characteristic of eukaryotic cells.

A variety of other differences in both structure and function also exist between the two types of cells. The upshot of these disparities is that scientists can perform their experiments much more easily upon the smaller, more comprehensible prokaryotes; but their observations do not necessarily hold true for the eukaryotes. With only the vaguest understanding of the structure of DNA in the chromosomes of higher forms of life, scientists are now working to bring their knowledge up to the level of sophistication they have achieved with prokaryotes. Because of the linear nature of the genome—linear messages are simple and direct; all the cell has to determine is where to start and when to stop—research is being undertaken to isolate individual eukaryotic genes and to determine how their function is affected by their neighboring genes. The method used to isolate each gene and have it express itself so that it can be studied is recombination. In fact, recombination is the only method yet devised that seems to permit the successful expression of less understood eukaryotic genes in prokaryotic hosts.

The First Fruits of Genetic Engineering: Medical Possibilities

Understanding more about the way genes function may, in the future, lead to a fundamental revolution in the way medicine is practiced. Until then, however, the techniques of recombining DNA molecules and other discoveries have tremendous potential in improving the sophistication of present medical treatment.

Possibly the most important and most immediate achievable product of recombinant DNA research lies in the production of many of the more critical medications needed for both maintenance and cure of sick patients. Because of the technique's capacity to purify and recombine single genes before transporting them into new hosts, bacteria modified by the addition of the gene which produces the protein hormone insulin, for example, could then be cultured and grown commercially, in huge quantities. The growth itself would take very little time; *E. coli*, for example, divides every twenty minutes, and with each division proliferates geometrically. At the end of the growth cycle, the insulin produced by the living bacteria would be isolated and purified, ready for use.

The bacterial production of human insulin would circumvent many of the problems that diabetic patients face today. Currently, the insulin used by diabetics is a purified mixture of insulin from the pancreases of cows and pigs slaughtered for food. The worldwide demand for the protein hormone has put a strain on supplies, and producers predict a time in the near future when the demand will exceed the amount they can manufacture. In addition, some diabetics develop allergic reactions to animal insulin or to the chemical precursors that producers are still unable to separate from it.

In fact, the technology needed to turn *E. coli* into tiny factories of insulin production is just around the corner. Already scientists have succeeded in engineering a bacterium which might produce insulin, by recombining the insulin gene from a rat with a plasmid and inserting the hybrid into *E. coli*. The hybrid has grown and has been copied by succeeding generations of *E. coli*. The next step is to determine how to persuade the transplanted genes to function and produce insulin in the alien host. When this has been accomplished, the minimal difference between the insulin-producing genes for rat and human cells—which are both eukaryotes—will make the transition from the production of rat insulin to the production of human insulin relatively simple. And the 2 million diabetics in the United States and 40 million around the world will have a safe, consistent supply of the one hormone they need to stay alive.

Success with insulin will have repercussions in the production of practically every organically derived medicine known to man. The clotting factor, a protein which protects normal people from excessive blood loss and which is missing from hemophiliacs as a result of a single genetic defect, also seems susceptible to production in small bacterial factories. And antibiotics, which already are isolated from the bacteria and fungi in which they grow naturally (except for the penicillins, which are synthetically constructed), can be introduced into more efficient, more prolific organisms, both increasing their availability and cutting their cost.

Recombination might also prove to be the answer to a problem that has been dogging scientists since antibiotics were first discovered: while antibiotics can successfully and safely treat bacterial illnesses, the only method now known for protecting organisms against viral diseases involves taking the highly volatile and potentially dangerous killed or attenuated viruses themselves and injecting them into the patients. Theoretically, the dead or weakened viruses can trigger the immune response of the body to produce the antibodies necessary to fight the disease. Practically, there is almost as much chance of the patient contracting the disease as there is of his producing antibodies.

Viruses are responsible for a vast smorgasbord of diseases, from the common cold to smallpox, from gastrointestinal discomfort to the deadly hemorrhagic fevers of Africa and South America. Finding a drug that can fight them has proved to be a hopeless task, mainly because viruses proliferate only within the normal cells of the body: it has been virtually impossible to attack the viruses without harming their hosts.

But scientists have learned that it is the virus' protein coat which contains the antigens that induce the body to produce its antibodies, while it is its DNA which causes the disease itself. Already, an *E. coli* plasmid coding for an enteric toxin fatal to livestock has been taken apart so that the toxic element could be separated from the rest. If the remaining parts of the gene—carrying the antigens— are grown and injected into cattle, it is highly probable that the animal's immunological system will construct the antibodies needed to defend against the disease. And if similar techniques are used to separate the toxic portions of a virus from its protein coat, it is also possible that human immunological systems can be taught to defend the body against viral diseases it has never before encountered. Scientists might also be able to isolate the antibodies themselves, grow them in an *E. coli* host, separate them from the host, and inject them wherever necessary to defend against disease.

Amniocentesis: The Prediction of Genetic Defects in Fetuses

A small amount of uterine fluid containing at least some fetal cells, a few weeks of culturing, and a look under a microscope—that is all that is now required if physicians wish to evaluate the chromosomes of a fetus for gross genetic defects. The method, called amniocentesis, was only recently developed and brought out of the realm of experiment and into everyday obstetric care.

Amniocentesis, by itself, would seem to have little to do

with genetic engineering; all it does is give the physician an opportunity to examine the chromosomes for Down's syndrome (mongolism), the sex of the unborn child, and more than sixty rare biochemical abnormalities. But its development is crucial to many of the speculative possibilities of genetic engineering, for it opens the door to a flood of other prenatal diagnostic techniques which presage the advent of genetic surgery and alteration of the unborn child.

So far, amniocentesis is used mainly as a means of identifying defects that might require selective abortion. Its accuracy (better than 99 percent) and safety are well known, but its future is still somewhat in doubt. Amniocentesis may well be the precursor of prenatal diagnosis of more and more subtle genetic diseases. Some may turn out to be so debilitating that abortion is the only solution; others may be susceptible to *in utero* chemical treatment and cure; still others may require immediate and unusual treatment in the hospital at the moment of birth. The ability to extract and culture cells from fetuses will undoubtedly lead to refinements of diagnostic techniques; whereas it is still only possible to examine the grosser structures of the chromosomes, it may someday be possible to evaluate the entire genetic complement of the infant for signs of abnormality, down to the level of its nucleotide sequences. Amniocentesis, beyond its usefulness as a technique, is a conceptual breakthrough as well; it enables scientists to envision the potentials of treatment at the source, at the levels of the reactions of protein synthesis and differentiation themselves.

Caution: Bacteria at Work

In the depths of a fierce Atlantic storm, an oil tanker runs aground. Its sides split open like weak seams in a too-tight dress, and it begins to pour oil into the surrounding water. The crew radios for assistance, and immediately two separate teams begin their crucial tasks: one leaves to save the men on board; the other works to contain the spill.

From Saudi Arabia, samples of the oil from the supertanker, drawn before it ever left port, are flown to the

United States and analyzed for their proportions of hydrocarbons. A helicopter hovers near the foundering ship, takes samples of the ocean, and returns them to the laboratory for analysis.

When the analysis is complete, both the polluting spill and its environment have been measured for contents, temperature, salinity, and impurities. The list, pages long, is brought to a bacterial factory nearby and handed to the scientist in charge. A novel bacterium, never before created and tailored to meet the exact specifications of that one spill, is about to be born.

The scientist takes the list and begins to plan the birth of the new bug. He is like an architect, designing a building to function in ways that buildings have never been able to function. Slowly, he maps out his strategy, planning the transduction into his bacteria of plasmids that will turn the oil's hydrocarbons into food for the organism, plasmids that will enable the bug to eat and reproduce at its optimal capacity in the cold, slimy waters of the Atlantic, plasmids that will ensure the death and breakdown of the organism when there is no oil left to consume.

When his blueprint is complete, he goes to the great walk-in freezer that houses huge vats of frozen, sealed plasmids and bacteria, and makes his selections. Then he moves everything into the vast, two-story production room, and, step by step, begins to build.

Soon he is finished. From harmless bacteria and homeless plasmids, he has created a tiny monster, capable of extraordinary devastation within the narrowest of parameters. He sends for the helicopter.

The helicopter attaches the now-sealed vat to its undercarriage and ferries it to the scene of the accident. Below it, the thick, black scum of the slick stretches for a mile beyond the tanker, floating with the currents. Tiny pinpoints of yellow in the water are attaching hoses from four other tankers that are attempting to siphon off and save as much of the oil as possible from the doomed tanker, now rolling on the rocks like a beached whale.

The helicopter hovers over the spills, then slowly lowers the vat into the water. It opens. Millions of microscopic organisms pour out, hungry for oil, ready to consume and divide and consume again. The helicopter retrieves the vat and ascends to a new height to watch. Soon a small path of clear water can be seen, approximating the outline of the vat. Slowly, ever so slowly, the edges of the clear patch begin to expand. It will take days, but the course of events is clear; the spill will be beaten.

The helicopter turns toward home, leaving the bacteria to do their work. When the spill is gone, they will die. The oil they have consumed and used as food will have become little more than a part of the bacteria, now food for the plankton that cover the ocean.

It may sound like science fiction, but it is not. General Electric's Dr. A. M. Chakrabarty recently fashioned the first crude prototype of these bacteria in a marvelous demonstration of the potentials of genetic engineering. And although the bacteria that Chakrabarty constructed are far from ready to be used in practical situations, the very fact of their existence is an indication of the degree of sophistication that genetic engineering can achieve.

Chakrabarty's method of creating his novel organisms was relatively simple. Oil is little more than a mixture of different kinds of hydrocarbons—aromatic, aliphatic, terpene, polynuclear aromatic, and cycloparaffin. And a variety of existing bacteria use different kinds of these hydrocarbons for food. In fact, one early technique of searching for oil underground was to find areas where a large number of hydrocarbon-consuming bacterial strains were concentrated.

Unfortunately, each strain of bacteria can degrade only one kind of hydrocarbon; each, alone, can attack only limited components of the oil. And simply throwing together a group of different bacteria, each of which degrades a different hydrocarbon, would not work because the presence of some bacteria adversely affects the survival of others. The inhibitory interactions among the many bacteria would kill all but a few varieties.

It occurred to Chakrabarty that he could solve the problem of adverse interaction if he could extract the hydrocarbon-consuming properties of each bacterium (without also extracting its toxic properties) and transfer them to a new, hybrid bacterium. Since the enzymes which act upon hydrocarbons are produced by plasmids, Chakrabarty had his own ready-made vehicles for transduction.

One by one he transferred the different properties of consumption to his hybrid. When he was finished, he had a bacterium that could consume four different types of hydrocarbons at the same time. In addition, it grew far faster than any of its ancestors had. And although he stopped adding new qualities to his hybrid after it had accepted four quite readily, he noticed that it had the capacity to accept several more.

Chakrabarty's work with hydrocarbon-consuming bacteria stopped soon thereafter, partly because of the outcry that followed word of his success. The potential for danger, in this instance, is obvious: if a bug is constructed that can consume oil, and if that bug escapes either from the laboratory or from the oil spill and insinuates itself into somebody's gas tank, the potential for an epidemic-size attack of oil-consuming microorganisms upon the machinery of the world, degrading every last drop of oil and bringing civilization to a dry, grinding halt, would seem to be tremendous.

The answers to this charge are strong but inconclusive. First, the nature of epidemics and the difficulties in getting one started probably preclude *any* epidemic from reaching a scale as large as the one envisioned. Second, Chakrabarty's bug was not a novel organism, at least not in the way that those concerned about recombinant research picture novel organisms. His bacteria simply hosted a number of naturally existing genetic factors that had been distributed among several varieties of bacteria. Those varieties exist all over the world, and have never been known to start an epidemic of oil degradation on any scale, even though each one has the capacity to sip off enough of the oil's potential to lower octane ratings below usable minimums. If the danger of a devastating epidemic does indeed exist with Chakrabarty's bug, it surely exists even now, when donors of his bug's components are floating loose around the world.

Contrary to a widely held misconception, Chakrabarty's work did not make use of the techniques of recombining DNA molecules. Although certain mechanical aspects were the same—for example, the transduction of plasmids

Now you can own it all.
Everything Shakespeare ever wrote. Every comedy. Every tragedy. Every historical drama. Every epic and romantic poem. Every finely-chiseled sonnet.

They're yours in three handsome volumes for only $1, as your introduction to The Classics Club.

The Classics Club is quite unlike any other book club. It doesn't offer best sellers that come and go. Instead, it offers its members a chance to stay young through great books that will never grow old.

These books include Utopia by Thomas More; the wisdom of Plato, Aristotle, and Marcus Aurelius; Benjamin Franklin's Autobiography; Omar Khay-

(Continued on other side)

into bacterial hosts—no potentially novel genetic sequence was constructed. Indeed, he did little more than artificially induce precisely the same kinds of plasmid transduction that takes place daily between bacteria in nature. But his creation of oil-eating bacteria—and his other attempts to transduce into bacteria the capacity to bind precious metals—follow a course parallel to that of research into recombination. Chakrabarty's work provides a fine indication that the more controversial scientific advancements do not proceed in a vacuum; they are, instead, intimately tied to other, equally sophisticated avenues of research. Like toothpaste that has been prematurely squeezed out of the tube, research that has been developed cannot be simply undone, for its very existence opens up possibilities in allied fields which could prove as beneficial or as harmful as the original research itself. This is one of the reasons that the controversy over recombinant DNA research must include discussion of the far more general problems of bioethics, and of the ways in which different avenues of research affect each other, in addition to the immediate concerns emanating from the biohazards of the basic research itself.

Nitrogen Fixation and the Growth of Crops

Although the problem with the world's food supply is one of distribution rather than quantity, we are fast approaching an era when, if we continue our present modes of production, real scarcities caused by inadequate supplies will become facts of life. To ward off that imminent catastrophe, scientists have been struggling with ways to increase yield, cut costs, and reduce the massive expenditure of energy now needed for agricultural production. While the "green revolution" has increased the yield of such basic crops as corn and rice, one way of the solution to all three problems may surface through the use of the techniques of genetic recombination.

Fixing nitrogen means taking nitrogen molecules, which are inert, and converting them to ammonia, which plants need for growth. Presently, the only technology that has been developed for supplying plants that do not have their own capacity to fix nitrogen with the ammonia they need involves using synthetic ammonia fertilizers that farmers dump, by hand or machine, onto their fields. Synthetic fertilizing is extraordinarily expensive, and time-consuming. Making them requires massive amounts of natural gas; spreading them requires huge chunks of the farmers' time.

Recombination leading to natural nitrogen fixation is based on the ability of several varieties of bacteria to perform the conversion themselves. These bacteria, which often live in a symbiotic relationship with legumes—one of the few kinds of plants that do not require ammonia fertilizers—live and grow in nodules projecting from the legumes' roots. From there, they leach nitrogen out of the soil, convert it to ammonia, and supply it to the plant.

Scientists see two possible ways of endowing other crops with the capacity to fix nitrogen. The first is to alter the genetic structure of the plants themselves, giving them the inherent capacity to catalyze the conversion. This method has several drawbacks, not the least of which is that scientists are not even close to discovering the methods needed to make sophisticated, delicate genetic surgery a reality, and will not be in the foreseeable future. In addition, it requires unusual amounts of energy to fix nitrogen. If the plant is required to supply its own energy, crop yields may actually decrease, since energy that would otherwise be used to produce other useful parts of the plant would have to be rechanneled.

The second method, and the one upon which scientists' attentions are really focused, is to transplant the genomes of bacteria which have the capacity to fix nitrogen into bacteria which act as symbionts to the crops that now require synthetic fertilizers. The so-called *nif* (nitrogen fixation) genes have been isolated, and the first attempts at recombination have already taken place.

Recombination is only one of several ways of achieving this second method of fixation. Mutation and transduction both have been shown to be successful in achieving at least some of the goals necessary for success. Nevertheless, whether or not recombination is the tool that leads to nitrogen fixation in plants, there are serious environmental questions that the potential technology raises. A major consideration lies in the production of excess ammonia.

Will increasing the ammonia production by organisms in the soil and water have a detrimental effect upon the environment? There are a number of ways this might occur.

First, ammonia is toxic; overproduction could increase the alkalinity of the soil and water, or might lead to the escape of poisonous ammonia gas into the atmosphere. In addition, excessive ammonia in the oceans might lead to faster entrapment and conversion of carbon dioxide, decreasing the mean temperature of the earth and leading to shorter growing seasons and fewer crops (and even a new ice age).

Overproduction of ammonia might also directly accelerate the depletion of ozone in the stratosphere; any nitrous oxide (a product of some soil bacteria's interaction with ammonia) that escapes to the upper atmosphere will, in turn, be converted to nitric oxide, a molecule that catalyzes the breakdown of ozone. Indeed, some investigators believe that synthetic ammonia fertilizers are even now a major cause of the atmospheric breakdown of ozone.

With success in realizing either bacterial or plant production of ammonia still so far off, both the benefits and the risks of the technology are largely hypothetical. Still, nitrogen fixation will be one of the major goals of biological research in the near future; once it is brought into production and put into practical use, its consequences, whether they are beneficial or harmful, are things we will have to live with.

The Impact of Genetics upon the Future

In speculating about the future course of genetic research, and in considering areas far beyond our present knowledge, it is easy to indulge in baseless fantasy, to toss predictions around like so many promises in a political campaign. At this point, we know so little: we are like fifteenth-century Europeans looking out upon the ocean, trying to decide whether the earth is as flat as all the evidence suggests, or round and inundated by water, as the dreamers believe, so that sailing ever westward would uncover a direct route to India. It is not that prediction of distant discoveries makes no sense, for curiosity and fan-

tasy have served as the stick and carrot for science from
the beginning. It is more that supportive evidence for advances in applied research must come from basic research,
a field that is itself still speculative. Thus, evidence about
the validity of our speculation itself rests upon speculation—a weak foundation supporting a cardboard house.

Nevertheless, to refrain from guessing at the future of
genetic research is both impossible and unwise. The possibilities themselves are so tantalizing that they have mes-
merized writers of science and science fiction from H. G.
Wells to Theodore Sturgeon, from Mary Shelley to Isaac
Asimov. Whole societies, too, have been swept along by
the dazzling and seemingly limitless possibilities of intervention in the genetic construct of their populations, from
the Spartans who exposed their sickly babies to the elements, to the Nazis' attempt to create a master race.

It is not difficult to see why. Imagine, for a moment,
what might be accomplished once we understand how
genes are controlled, once we know how we develop from
a single cell into an immeasurably complex adult: genetic
surgery, which might be able to delete defective genes,
block abnormal ones, or add new ones if someone is
caught short; control of the inheritance of babies, ensuring
that only those with ideal genetic complements are permitted to be brought into a world already overburdened, and
condemning those whose genes do not achieve the norm to
lives of sterility—a policy, in other words, of active eugenics; cloning in animals, so that the strongest, gentlest,
meatiest livestock are born and born again, ensuring a
plentiful supply of food; cloning in humans and the establishment of armies of perfect soldiers bred for fighting,
armies of workers designed for manual labor, armies of
teachers, administrators, musicians, scientists, leaders, fol-
lowers; new forms of fantastically destructive biological
weapons, some which might sow defective genes among
the enemy to weaken his armies, others which might modify the bacteria in his fields, destroying his food crops.

From our present, limited perspective, all this and more
is possible. The technological advances—in cloning, genetic surgery, biological warfare—either have been or will
soon be attempted in laboratories in their crudest, most
primitive forms. Those potential tools of governmental

policy—eugenics and the weapons of biological warfare—
are bound to be used in more or less limited forms once
the hardware for making them practical is developed.

The question we must begin to ask now is not whether
scientists will attempt the experiments that will lead to
these fantastic results—if we permit it, they will—nor
whether they will succeed—we cannot know—but
whether, if they succeed, the results will be more likely to
strengthen us or destroy us. To answer this question requires more than a simple balancing of intuitive guesses
about risks and benefits, although determining whether the
future technology could actually be used detrimentally is
of crucial importance. Sooner or later we are going to
have to ask hard questions of society as well, to determine
whether it is mature and healthy enough to handle such
awesome power, or whether it must sometimes be protected against itself.

But these important questions come later. First we must
use our meager knowledge to evaluate some of the technological prospects in the future of genetic research. While it
is true that we begin with few facts, delaying the start of
the decision-making process until all the evidence is in is
waiting until the technology is imminent. At that point,
the genie is already out of the bottle; and, once it is out, it
cannot be stuffed back in.

Genetic Surgery

As many as 5 percent of the population in the United
States—one out of twenty people—are born with genetic
defects. While some defects are only slightly disabling or
have been shown to be susceptible to treatment, others,
like Huntington's chorea, sickle-cell anemia, and hemophilia, have either resisted treatment or require medication
too rare and too expensive to have widespread applicabil-
ity.

The recent surge in our understanding of genes suggests
that certain forms of genetic surgery are imminent.
Indeed, genetic surgery, in the guise of recombination,
does work in prokaryotic cells, by permitting the transfer
of tiny genetic subunits from one organism to another to
alter the function of the new host. Although splicing a

gene into a plasmid of a single-celled bacterium is an entirely different thing from altering the sequence of nucleotides on the chromosomes of a billion-celled mammal, the latter possibility does exist.

One sure division between what is likely and what is not lies between those defects that are monogenic—caused by one gene, or even one misplaced nucleotide in one codon—and those that are polygenic. While each gene produces a single substance, that substance may be only one factor in a long series that combine to create a single, visible characteristic; many genes may have to interact to produce a given result. The process is analogous to the complexity of trying to route a train through a freight yard: every switch on the way to the train's destination must be set correctly, or it will not arrive. Furthermore, switches, like genes, may take part in the movement of several trains at once. If a faulty switch is simply fixed in a position that permits the smooth routing of one particular train, its immobility might preclude the normal operation of other trains.

So it is with genes. Often they are each merely one step in a series of steps; often they can affect more than one process at a time. Unless scientists know the precise capacities of both gene and reaction, genetic surgery is as dangerous as trying to wire plastic explosives blindfolded. For the patient, at least, the results will probably be quite similar.

Polygenic defects, therefore, are probably beyond the range of genetic surgery as it is presently imagined. The variables are too great, the risks too obvious. But diseases like sickle-cell anemia and hemophilia are caused by monogenic defects. In both, the alteration of one nucleotide—and thus one amino acid and one protein—results in disease.

Although the problem of genetic surgery to correct monogenic defects is at least theoretically manageable, one immense roadblock still remains. Scientists have demonstrated the ability to alter the genetic composition of bacteria. But purposefully manipulating the nucleotide sequences in an organism composed of eukaryotic cells requires finding a way to get the right nucleotide to the right place in the right cells at the right time.

The steps necessary to meet each of these criteria have not only not yet been attempted; they have not yet been imagined. First, scientists must locate, for each disease, every organ or cell group containing the active defective gene. Then they must devise a method by which every one of those cells—which may number in the millions—can not only pick up the new genetic information but express it as a replacement of the old, defective information. It will take more than microscopic lasers or other hardware to succeed in this step; it will take a kind of surgery in which the surgeon's instruments are biological: tiny viruses, coded to intervene at the precise point of the problem, myriad enzymes to open each cell to the viruses, to cut the cell's existing DNA, to replace it with the modified version carried by the virus, and to anneal the rearranged molecule without leaving a scar.

Some of these diseases manifest themselves in adults, and will be relatively easy to treat. With the patient in front of him, the surgeon might, for example, excise the diseased tissue, manipulate it genetically in culture, and return it to its rightful place (even removing and returning tissue in this way will require techniques of grafting on a scale that does not yet exist). But what if the patient is a fetus, diagnosed through amniocentesis; what if that fetus must be treated in the womb to survive? What of such diseases as phenylketonuria (PKU), which, while it can now be treated prenatally through the mother's diet, cannot yet be cured, and is simply passed on from generation to generation? Or Down's syndrome (mongolism), in which each cell has three chromosomes in its twenty-first pair instead of two? Diseases like this must be cured not simply in the prenatal stage, but practically at the moment of conception, when the embryo is still small enough to permit surgery on each of its few cells.

Benign uses of genetic surgery might have other applications as well. Living organisms are genetically timed to degenerate. "Those not busy being born are busy dying" is not just a rock-and-roll, pseudo-psychological judgment; it is a genetic fact. Scientists have begun to realize that damage to the DNA molecules or cross-linking of the DNA in cells, which occurs throughout our lifetime, may actually prevent the mRNA from accurately "reading" the genetic information. As a result, cells produce inactive, inefficient, or even nonfunctional enzymes which inhibit our ability to maintain the body and its functions.

Genetic surgery, and the recombinant techniques that may well make it possible, is one way—perhaps the only way—of understanding the problems of aging and degenerative diseases. New techniques may make it possible to slow down or even reverse the aging process by repairing damaged DNA molecules or by preventing their cross-linkage. Already preliminary experiments have been tested under laboratory conditions. If the research succeeds—and if man is able to control the system of regulation that conducts the decline of an organism as well as its growth—genetic engineering could well provide us with the capacity for immortality.

But as bright as the possible benefits are, the potential for misuse is just as great. A virus that can infect and alter the genetic complement of an individual can almost as easily be made to perform a similar function upon a population. And control of genetic mechanisms that cause monogenic defects may lead to control of those involving polygenic defects as well—leaving such indefinable, complex qualities as intelligence and emotion (which, moreover, are often products of a combination of genetic and environmental influence) open to genetic tampering. The possibilities on both sides are enormous. Once again the question of control falls on the shoulders of society.

Cloning

Cloning is, quite simply, the creation of a group of cells or organisms from a single individual by means of asexual reproduction. It is a natural means of reproduction for everything from single-celled organisms to various groups of plants. It is also a natural part of human reproduction; since .5 percent of all live births around the world result in identical twins, triplets, or other multiples, over 200 million people alive today are actually clones.

The question raised by genetic engineering is whether we can devise a system whereby we can control the cloning process, manipulating the potential of cloning for our own good. The second question, an ethical one, concerns whether the risk of abuse of the power to clone will ultimately outweigh its potential benefits.

Certain types of cloning have already taken place. In 1965, J. B. Gurdon of Great Britain managed to clone the African clawed frog, in an experiment that has been repeated numerous times. To do so, Gurdon took a fertilized frog egg, irradiated it to destroy its nucleus, and surgically transplanted the nucleus from the cell of a tadpole into it. The process was complex and inefficient; only a few of Gurdon's clones grew and lived, and none actually reached mature adulthood. But his work showed that cloning is possible in at least some higher forms of life.

Whether or not Gurdon's discoveries presage cloning in mammals is another matter. A frog's egg is several millimeters in diameter, a comparatively huge target for the genetic surgeon's knife. The human ovum is microscopic, tiny by comparison; to perform similar surgery upon it would require tools and procedures not yet devised.

Another question raised by Gurdon's success concerns possible differences between the characteristics of amphibian and mammalian genes. By demanding that the genetic complement of an adult, transferred to a single cell, perform all the processes necessary to create a new organism, cloning requires that the adult cell revert back to its primary, undifferentiated, unsuppressed state. It is quite possible that there is something that exists only in amphibians which allows a normal cell to be switched back to the beginning of its evolution. Mammalian cells, which seem to become differentiated in the earliest stages of embryonic growth, may contain processes that almost irreversibly differentiate cells and suppress genes, making reversal of the process in a potential clone extremely difficult. (Recently, a purported case history of the first successful cloning of a human being was published,[1] amid furor in both the scientific and lay communities. This book raises dozens of ethical problems all by itself; some involve the questions of whether human cloning is a fit subject for scientific inquiry, while others revolve around the method that the author chose to disclose his claim—a method which permits no serious scientific discussion of the cloning and which is bound to make a bundle of money for the writer. Unfortu-

[1] David Rorvik, *In His Image* (New York: Lippincott, 1978).

nately, the account itself contains little or no scientific documentation for its most revelatory claims: the problem of dedifferentiation, for example, is glossed over in two or three sentences, and the only proof the author is able to offer that the child is actually a clone is the word of the millionaire who paid for the experiment. As *The New York Times Book Review* aptly noted, "Until further enlightenment [Rorvik's] book must be classed as quasi-science and placed with the books of Erich von Däniken and the authors of the Bermuda Triangle school."[2])

While the nature of the regulatory mechanisms along the chromosomes might prevent us from cloning copies of adults, it does not preclude our taking the embryo before differentiation occurs, tampering with the first few divisions of cells, and creating clones of as yet unborn infants. The drawback, of course, is that, in cloning an embryo, we would have no idea which of the practically infinite genetic paths the embryos might take unless we had mapped beforehand the entire set of nucleotide sequences in human beings—several billion units long—and traced all the possible alternative arrangements. Today, geneticists are still struggling to map the relatively few genes on some tiny viruses, several bacteria, and the fruit fly, and are years from even attempting to study the sequences of human genes. (Similar problems would not necessarily arise in the cloning of livestock, since careful breeding over thousands of years has led to animals with relatively few genetic alternatives; breeding a muscular bull with a heavy-set cow and cloning copies of the fertilized egg would, in all probability, lead to meaty, heavy-set offspring.)

If we do not have the capacity to clone adults, the fantasies surrounding the cloning of human beings fall apart. The chief advantage to cloning would lie in our ability to make copies of known quantities, but there is no way to discern whether someone will turn out to be an Albert Einstein or an Albert de Salvo simply by studying an embryo. Furthermore, even if it were possible to clone adults, it would probably be an impractical tool, superseded by sophisticated techniques of psychosurgery, which, by the

[2] *The New York Times Book Review*, May 7, 1978, p. 46.

time cloning becomes a reality, will probably be able to turn the meekest men into tigers, scatterbrains into mathematicians, and wallflowers into Don Juans. Cloning an army of soldiers would still take years of time and energy as the legions grew to maturity. Governments would be forced to plan their wars twenty years in advance to avoid the complete waste of cultivating an army that has nobody to fight. Far simpler, more practical solutions lie in the probable results of the new research in neurobiology, the study of the brain, which will undoubtedly pinpoint centers of aggression, centers of growth, centers of memory and obedience. Sooner or later, somebody will figure out how to temporarily activate the series of locations on the brain that will turn out soldiers as brave and fierce as any clones, workers as durable, teachers as incisive. Then, when the urgent need arises for a certain kind of man, volunteers will simply queue up at their local psychosurgery centers, undergo cerebral stimulation, use their newfound powers to rectify whatever emergency exists, and return to the centers for reversion to their original states.

If biology is in the forefront of the scientific revolution, and if genetics is the vanguard of the new biology, recombination is the spearhead of genetics. The once awesome mysteries of inheritance have been pulled apart, examined, and reduced to secrets, the secrets into sets of translatable codes. Now we are constructing the means of deciphering those codes in the hope that our ability to understand them will turn them from obstacles into weapons, from the proverbial brick wall against which we have been butting our heads into foundations of a new technology, one that will help us conquer many of the limitations that have kept us so vulnerable. Unlocking the secrets of genetics means opening the vault of nature, understanding how and why living things function and die, and discovering ways to turn the processes of life and death to our own advantage.

At the same time, every hypothetical benefit of the new technology—in recombination and in other forms of directed genetic manipulation—carries its own mirror image of risk. As in a chess game, where every move, no matter how powerful, contains elements of both weakness and strength, these new scientific tools contain the seeds of both our elevation and our extinction. And while we may become exhilarated by their potential for good, we must also soberly assess their risks. Choosing to view only one side of this issue is like choosing to become blind. And being blind has little to offer us beyond a heightened propensity for walking off cliffs.

FOUR:

The Risks

Despite the emotions generated by the vast potentials and risks of recombinant DNA research, the scope of the debate seems to have somehow outstripped the issues. Aside from the scenarios of horror that have been painted by the opponents of the research, aside from the very real dangers that might arise from our application of new knowledge, aside from the threat of accidents, there seems to be something more, something inherent to the scientific process that has led to many of the peripheral questions that have surfaced.

The key to this scientific puzzle is in the unpredictability of research itself. From a distance, the course of research seems linear. It is easy to see how one discovery leads to another—how, for example, the insights of Oswald Avery and Erwin Chargaff were crucial to Crick and Watson's final, correct model of the double helix. And while a quality of trial and error does come across in the story of their discovery, that feeling is muted in the light of more recent research, where the actual process of discovery seems to be so much more scientific, so much more logical and exact.

But that feeling of inevitability is precisely where the illusion lies when those of us who are not directly involved in research try to examine the process. For as precise as science has become, the nature of its actual discoveries has become even more complex, even more fundamental, even more unpredictable. Research does not simply march straight ahead to its logical conclusions. More often, it stumbles around, attempting to pick a safe path about as successfully as a drunk trying to negotiate an icy road during an earthquake. The unrewarded search for a correct solution to a problem is as much a part of science as progress, and the failure of a project may turn out to be as important as its success; in the end, discovering what is not true about something may well clear the way for someone else's insight into what is true. Laymen and scientists who are not intimately involved in the details of a specific line of research seldom hear of the hundreds of failures that precede success. For that reason, their judgment of the process may be jaundiced, based upon a perception filtered through the little they know, and leading to conclusions founded upon the most superficial of appearances.

It is also clear that the ramifications of a specific piece of scientific research do not simply end with a successful discovery. It is almost impossible to trace the future of most scientific discoveries and the possible uses that may be found for them. Questions that might have been asked about past creations now seem obvious in the clear light of history: Would the chemists who so enthusiastically invented the technology of oil and its petrochemical transformations have pushed ahead so readily if they had been given a whiff of the air we are now forced to breathe? Or would the physicists who worked with nuclear fission have been so overjoyed at their success if they had been permitted to see future generations living under the threat of the annihilation of civilization? (Even though it is true that Otto Hahn, who first split the atom, was said to have cried out when he recognized the power of his discovery: "But God would not have wanted this!")

These are questions that we obviously cannot answer. Yet they are relevant to our present discussion, for similar problems are bound to crop up with the products of recombinant DNA research. Beyond the puzzles of the basic research—and the horrific-sounding accidents which may or may not occur—lie the probable burdens that the technologies generated by the research will place upon us, the acknowledged impact that recombination will have upon life as we know it. The new truth embodied in recombination—in the controlled rearrangement of genetic material—is relatively simple; but its future implications are so complex and so unforeseeable that the mere fact that most scientists see nothing dangerous in it is not enough; what seems safe can become catastrophic with or without an observer's ability to recognize its potential.

So we are faced with yet another paradox. The risks are unforeseeable. Yet if we do not attempt to counter them before they arise, we are derelict in our duty to the world and civilization in which we live. If we refuse to search for something that we know we cannot find, we condemn the future to whatever the new technologies have to offer. We are therefore obligated to push our study despite the odds, to try to predict so that our preparations in themselves become barriers to the unknowable.

So what are the risks? Fundamentally, they take two forms: those which may occur during the process of basic research—classified mainly as potential "accidents"—and those which may surface when discoveries are applied, when we are actually using the technology that follows them.

The risks of basic research are relatively clear. Because the process of recombination is straightforward and well understood, the dangers can be mapped. First, there is the possibility that bacteria hosting the recombined, hybrid DNA will become altered in such a way that they gain a greater capacity to cause disease or resist treatment. It is generally agreed that the genetic determinants capable of promoting certain kinds of pathogenicity do occur in the plasmids that are currently being used to introduce and transfer the new genes into bacteria; small tumor viruses like SV40 have been shown to cause cancer—probably not in man, but certainly in other animals. Furthermore, antibiotic resistance—a particular burden for science, now that the use of antibiotics is almost indiscriminate on a worldwide scale—is a major concern; geneticists' use of the genes of antibiotic resistance as markers in their experiments may well permit toxic bacteria not presently resistant to inherit that capacity. Already the usefulness of many drugs is threatened. Recombination might add to that list.

A second clearcut danger involves the transformation of relatively harmless strains of bacteria into those able to produce diseases like diphtheria or cholera, or toxins as deadly as botulinus. As the bacterium of choice in most experiments, *E. coli* has shown itself to be amenable to many biological transformations. But while it is the best-understood of the bacteria, its entire genetic potential is a long way from being completely mapped. And it, too, has its virulent strains. The possibility of interaction between the resident genes of experimental *E. coli* and newly acquired recombinant hybrids is not yet well understood.

Finally, as bacteria pick up the genes that are under study through recombination, there is no guarantee that they will not also pick up genes still unidentified. As the tiny virus Phi X-174 showed, genes can overlap, signals are not always simply linear in nature, and the introduction of one characteristic from a chromosome into a new host might carry with it the stop-start mechanisms of other genes. Methods of purification that try to ensure that only the desired genes are transferred are, in reality, only accurate to about 90 percent of purity; and while the last 10 percent can often be guessed at, its true composition is seldom known. Some recombinant experiments, the so-called "shotgun" experiments, do not even attempt to purify the DNA that is to be recombined: an entire complement of genes is shoved into the mix, hybrid plasmids are formed and introduced into new hosts, and only then, while they are growing, are desired characteristics separated from the rest. The shotgun method works on the principle that cream rises to the top, that the genes to be studied can be easily separated from the rest by various experimental procedures that kill whatever is not wanted. It is, for example, a favored method for experiments that require the use of antibiotic markers—for what could be easier than introducing into a new mob of recombined bacteria an antibiotic which will destroy all those that have not picked up resistance during recombination? The problem is that nothing is absolute; something previously unmapped in the shotgunned genetic complement, a quick transfer of the ability to resist from one bacterium to another, a quirk of experimentation, could all lead to the survival and possible growth of undesired, undetected, dangerous bacteria.

While it is relatively easy to pinpoint the major risks inherent in basic research, delineation of the risks of its application is far more vague, principally because we do not yet really know how the research will be applied. Nevertheless, the risks can be divided into three broad categories: discoveries which might seem useful or harmless in the eyes of both science and society, but may have unforeseen consequences; aberrant research that may lead to new biological forms that we do not want at all; and research that may seem fascinating and elegant to the scientific community—with possible applications to man's most basic environmental and medical problems—yet that remains unacceptable to society because its risks match or exceed its potential. But the issues of application cannot be dealt with prematurely; if we find that the basic research itself is so dangerous that it should not be continued, questions about whether or not the technology would be safe become irrelevant.

The end of the Asilomar Conference in February 1975 marked a point of no return for the controversy. The press' interest (and, with it, the public's) had finally been pricked; whereas the Berg letter and subsequent developments had resulted in frivolous headlines and stories, everyone could see that science and society were now being faced with a real issue, one that would not succumb to whitewash in scientific conferences or emotionalism in newsprint. An issue that could breed in-depth reporting in newspapers, magazines, and journals as diverse as *Rolling Stone, Science,* and the Washington *Post* had more substance than the cartoonlike extremes conjured up by some of the antagonists. The issue was more than a simple internecine struggle between professionals, more than a question of technique or jealousy. In fact, it was more important than almost any issue that had ever confronted science and society, for it marked the point at which an aroused public began to respond to the warnings of its watchdogs.

There is no question that similar widespread concerns should have arisen (and, in some limited strata, did arise) over the conundra of atomic weaponry, pollution, and the use of humans in scientific experimentation. But, in the psychology of the public, neither the time nor the issue fit the mood. We still trusted most of our institutions then; we still believed in the relative infallibility of our leaders. Vietnams and Watergates percolated around the edges of our minds, but never struck us as symptoms of our own societal conduct. Small, isolated groups criticized, marched, and turned harsh, cold spotlights upon shortcomings expressed by the people we had elected, while the rest of the population equated the "country" in "My country, right or wrong" with the people in charge, confused the office with the man, and showed disapproval of dissent with a clenched fist.

And then, quite suddenly, in the midst of *the* Vietnam, *the* Watergate, more people began to see that these were just men up there, people as blemished as the rest of us, who could stumble over the pronunciation of easy words, spew obscenities in private while preaching the morality of Mount Olympus, and cheat on their wives. And the little lightbulb called "idea" lit up above enough heads to make a difference: we realized that they are people like us, they can make mistakes as blatant as the ones we make. And with as much control as they have—and, perhaps, need to have, to run things smoothly—their mistakes might prove to be a lot more costly than our own. The shock waves rolled from spectrum to spectrum, through politics, economics, and professional conduct, through the educational system, the military, and, ultimately, through science.

And just as the first tremors began, an issue surfaced, one that would not go away, one that, unlike the bomb or bad water, had ramifications that could not be halted if everything went wrong. The other problems spawned by the meeting of science and society could always be handled, changed: we could stop building bombs if the need arose and throw the ones we already had deep into the six-mile trenches of the Pacific Ocean; we could close down the mills and factories that made some rivers deadly to any living thing, build catalytic converters for cars, and cut back on our systematic destruction of our environment. But recombination—that was a different story. Aside from the fundamental horrors dredged up by dreams of how we could use the new discoveries, here was something that seemed to present grave risks simply in the conduct of the basic research itself. And the dangers, once loosed, might not even be reversible. Some unnamed, potentially deadly bug, some undefined horror, could be set free in the world without our even wanting it to, without our having the first idea on how to recall or stop it. This was definitely something to be concerned about.

And so, after Asilomar, the issue of recombination was hoisted another notch toward the ultimate in recognition, Congressional action. What had begun as an informal dialogue between scientists and had become the subject of a widely publicized scientific meeting was now being placed firmly in the laps of the federal agencies that had funded the research in the beginning and would have to give their approval before financing could continue. Indeed, part of Asilomar's final statement included a charge to the National Institutes of Health. They were asked to investigate the current state of knowledge and technology regarding DNA recombinants, their survival in nature, and their ability to transfer their properties to other organisms, to assess the possibility of the spread of specific DNA recombinants and the possible hazards to the public health and the environment, and to recommend a set of guidelines to be drawn up on the basis of this investigation. As was later pointed out in one of the NIH's hearings on the subject, their most basic problem was "to construct guidelines that allow the promise of the methodology to be realized while advocating the considerable caution that is demanded by what we [the NIH's advisory committee] and all others view as potential hazards."

The advisory committee appointed by the NIH's director, Dr. Donald Fredrickson, was, at first, composed entirely of scientists; only later were spokesmen capable of voicing the public's fears brought in, and even they were some of the weakest available—a partial result of the authorities' defining "responsible spokesmen" as people who thought much the way they did. Nevertheless, the months that it took the committee to devise and construct its first set of guidelines—from April 1975 to June 1976—were packed with strong debate, powerful emotions, and extraordinary displays of scientific pique. The guidelines were drawn and redrawn three times in the first six months, concluding with a set agreed upon in La Jolla, California, in December 1975. The battles that preceded La Jolla were fierce; the preliminary set of guidelines, which bore a remarkable resemblance to the final recommendations of Asilomar, was first badly diluted, then strengthened in the face of mounting protests from scientists and laymen alike. By the time La Jolla rolled around, the advisory committee was faced with the realization that if it did not finally agree, it would have to admit a dismal failure; under that threat, the proposed guidelines were painstakingly hammered out and approved, one at a time.

In February 1976, the NIH was ready to hold a hearing to examine its proposed guidelines. There, in two days and in a remarkably open and frank atmosphere, the guidelines were picked apart by spokesmen from every possible persuasion, then slowly put back together. By June, they were finally ready for public consumption.

The guidelines received predictable reactions from the two sides of the issue. Although some researchers decried the relative constrictions on what experiments could be conducted and where, most breathed relief, recognizing that the guidelines permitted research to continue. In fact, after the months of work, the guidelines still looked like the Asilomar statement, swollen with elaborate detail and specific in some cases almost to a fault. Still, most scientists proclaimed the proceedings a huge success; the public had been consulted, science was being allowed to proceed, and the risks of the research had been dealt with, it seemed to them, eminently fairly.

But opponents of the research were not so certain. They quickly noted holes in the guidelines—which applied only to federally funded research, and said not a word about private enterprise and industry. And they did not hesitate to point out the contradiction between what the NIH was supposed to have done, and what it actually did:

According to the charge [of Asilomar]: first, research to assess hazards; second, and on that basis, the drafting of guidelines. It was indeed a rational order of action. Yet it has had to be reproposed—to no avail—by critics. For a curious thing has happened on the way to Asilomar and on the long road of deliberation travelled since then by the NIH committee. Guidelines have appeared, but research aimed specifically at defining the hazards is only just beginning. Furthermore, no restriction on the number of facilities engaged in this work has been set. Somewhere along the road, the yellow light for cautious research turned green for its proliferation.

Advocates of the present policy maintain that the public has participated in its formation. Let it be clear that expression of views to decision-makers is a quite different matter from participation in decisions. Through the mechanism of a technical committee, decision-making power has been concentrated in the hands of front-rank researchers, all of whom are committed to biological research in general and many, to recombinant research in particular. There has been no representation from those immediately at risk—technicians and maintenance personnel, for example; no representation from public interest and environmental organizations; no representation from the public at large.[1]

The objections voiced by Dr. Wright and others were indeed valid; the method of deliberations chosen by the committee was geared to the precariousness of the politics surrounding DNA. The scientists involved in the research generally agreed that some form of guidelines would have to be drawn up. They also felt that those which would be agreed upon had to be stronger than they thought necessary, for the simple reason that the public's advocates would not stand for anything less. But the continued focus on a search for technical solutions avoided for the second time the larger problems—those involving the fundamental validity of any research that so profoundly affects our existence. And the final guidelines, while relatively stringent, were strong for political, rather than scientific, reasons. The result was that few were satisfied with what the NIH published. Those actively concerned recognized that they were being presented with a paper target, signifi-

[1] Susan Wright, "Doubts over Genetic Engineering Controls," *New Scientist*, December 2, 1976.

cant but false. What the guidelines discussed was, of
course, important; it was crucial that some rules be set up
for the conduct of experimentation that contained such a
diversity of unknowns. But what they did not touch upon
will ultimately prove far more important. For the basic
questions of science's responsibilities to society—and society's to science—were really what the debate was one day
going to be about.

What the guidelines did encompass were the problems
inherent in basic research, the danger of accidents releas-
ing into the environment novel organisms that could neither be constrained nor destroyed. And, even there, the
guidelines were caught between inherent discrepancies.

The Problems of Physical Containment

Of the two types of containment called for by the
guidelines—physical and biological—physical containment
was to act as the first line of defense. In a structure similar
to that handed down by Asilomar, physical containment
consisted of four categories: P1 (minimal containment, requiring only the safeguards of good laboratory practice),
P2 (minimum isolation), P3 (moderate isolation), and P4
(maximum isolation). But while concepts like maximum
isolation look beautiful on paper, they fail to impress in
practice. The construction of a P4 facility is laudable. It
includes complete isolation for the organisms under experimentation, air locks, required showers for all personnel,
double-door autoclaves to sterilize all equipment leaving
the laboratory, separate ventilation and waste-treatment
systems that can sterilize air and garbage before it is sent
out, and zealous restrictions on who is allowed to enter the
facility and who is not. But it has been proved in practice
that no degree of physical containment is foolproof; it is
only as good as the people who are working inside it. Like
a perfect vacuum or absolute zero, it simply cannot be
achieved. As one of the principals in the debate said, "The
only type of [physical] containment that is worth anything . . . is the concern of the investigator involved. You
can sign all the permits in the world, and you can have all
the facilities . . . but if the people involved in doing the

work aren't worried about what they're doing and aren't careful, it's meaningless."

Even if the principal investigator is conscientious, he and his assistants are not the only ones with access to even the laboratories with highest security. Other personnel—maintenance workers, for example—move through the outer areas surrounding the restricted laboratories with relative impunity. And the obvious problems of less careful assistants washing, for example, contaminated hypodermic needles, puncturing the rubber gloves that they wear, and becoming infected is neither unthinkable nor unknown. (In fact, the only known death from cancer to be caused by a laboratory accident occurred in precisely that way in 1926, when Henri Dadon, a French medical student, pricked his hand with a syringe contaminated by a cancer patient; he developed nodules soon thereafter, and died a year later from metastasized tumors.)

There is, then, a human problem. And the guidelines, while taking note of it, did nothing to try to solve it. Neither did they discuss in detail the problems of mechanical failure. Although there have never been any known infections arising from organisms trapped in P4 sealed-glove cabinets (the totally enclosed cabinets that prevent any researcher-organism contact, except through the barrier of heavy rubber gloves), the cabinets have been known to explode accidentally, as one of those containing the first rocks to be brought back from the moon did. Had any tiny moonling organisms actually existed, they might have been blown all over the laboratory, the investigators, and, ultimately, the world. Other mechanical dilemmas include that posed by the autoclave and the nature of DNA. Although the autoclave is a powerful sterilizer, one liter of culture containing recombinant DNA might hold as many as 10^{14}—100 trillion—molecules. Since the recombined molecule is a strong, closed, doughnutlike structure, small amounts may, in such large quantities, have the ability to survive and to infect other cells (although recent experiments by Ronald Davis of Stanford University have indicated that smaller quantities of DNA are totally fragmented by autoclaving). Other, more likely possibilities include infestations of insects and vermin; cockroaches have been found within high-containment (P3) areas of the otherwise impeccable facilities of the NIH itself, and the Egyptian ants that haunt the laboratories of Harvard University have become famous for their ability to survive every kind of poison that has been thrown at them.

The skepticism which seems to greet proposals of physical containment is not only supported by conjecture. In January 1976, a report entitled "The Detrick Experience as a Guide to the Probable Efficacy of P4 Microbiological Containment Facilities for Studies on Microbial Recombinant DNA Molecules," by A. G. Wedum, was published under the sponsorship of the National Cancer Institute. Dr. Wedum's report surveyed the history of the maximum-containment facilities at Fort Detrick, Maryland, which, in the 1950s and 1960s, was the U. S. Army's center for research into the potentials of biological warfare. In the twenty-five years he surveyed, Wedum found evidence of 423 infections. Although there was a small decrease in their number as the efficacy of the mechanical systems increased, the only significant decrease in incidence of infection followed the introduction and use of live vaccines against the diseases being studied. Without the vaccines, and in experiments involving the more dangerous organisms, "there will be some laboratory infections, usually due to breaks in rubber gloves [attached to the safety cabinets], human exposure during entrance and exit of materials, leaks in the system, and human error."

The present state of recombinant research includes two alarming departures from situations analyzed by Dr. Wedum. First, no live vaccines exist for E. coli used in recombination—and even if one did, it would ultimately prove ineffective, due to the extreme proliferation of new, possibly resistant strains of E. coli that various laboratories practicing recombination would construct. Second, Dr. Wedum was examining practices and facilities at a single laboratory, one upon which the concern of the entire military complex was focused. Those people knew they were playing with fire when they handled organisms responsible for everything from Q and typhoid fevers to lymphocytic choriomeningitis and hepatitis. And still there were infections, caused by both human and mechanical failings. Recombinant DNA research, on the other hand, will be spread out all over the map. Some of the laboratories will be run as carefully as Fort Detrick was; some will not. Some will have all their facilities in working order; some will not. Some employees will recognize the seriousness of handling toxic agents; some will not. What is certain is that accidents will happen. Events will contrive to break through the barriers of physical containment, and novel organisms—whether they are dangerous or not, whether they can survive in the outside world or not—will be released.

As for the use of *E. coli* in recombinant experiments, Dr. Wedum concludes that the most effective safety measure to prevent infection of laboratory personnel would be to use microorganisms that cannot infect humans. The fact that *E. coli* can be transferred from person to person only through the mouth is both a blessing and a curse: because its methods of infection are so limited, laboratories will have fewer actions to safeguard against; but because of our propensity for bringing objects to our mouths, "it is probable that the most important of all precautions would be the hardest to enforce, namely (1) use mechanical pipettors, (2) avoid hand contamination, (3) keep hands out of the mouth and off of the face, (4) no smoking in the laboratory, (5) no candy, food, drinks, or gum in the laboratory."

Three other important points relate to the issue of physical containment:

- Physical containment functions much more effectively in situations where the dangerous material can be detected when it escapes—in laboratories working with radioactivity, for example, where there is an immediate, visible reaction in the change of color of radiation badges upon the release of excessive amounts of radiation. Recombination provides no such easy alarms. When *E. coli*, a human symbiont, escapes, the infection picked up by laboratory personnel may well be subclinical—that is, undetectable by either the individual infected or the laboratory set-up for monitoring his blood. Furthermore, laboratory personnel are likely to become more lax in their observance of safety procedures when they are dealing with a potentially infectious material whose escape they can detect by neither sight nor smell.

- P4 laboratories may prevent many leaks and accidents, but not all. And they are extraordinarily expensive—costing as much as $500,000 per laboratory. Working in them under the powerful hindrances of sealed cabinets, rubber gloves, mandatory showers, and the like makes some kinds of experimentation so tortuous that spills become actually more likely than they might be under P1 or P2 conditions. Furthermore, the expense and difficulties involved have already caused some researchers to circumvent the guidelines; they simply devise an experiment, determine whether it can or cannot be performed in their laboratories under the strictures of the guidelines, and then think up ways to circumvent the rules, conforming with them to the letter, but not in spirit. In some laboratories, spirit is a matter of personal interpretation.

- The difficulty of finding adequate P4 laboratories and of performing experiments in them makes many of the more elegant, more ambitious attempts at recombination impossible.

So it is that physical containment, while necessary, is deceptive. Far from being a panacea for the problems of novel organisms that escape, it has acted at least partially as a good pair of blinders, giving those in favor of continued unrestricted experimentation a means to shade the argument from the truth. Physical containment is in fact little more than a psychological protection against infection if it is used alone, for it provides nothing more than a means by which escapes can be postponed. And even that postponement is a delay only in mathematical probability; Murphy's Law, which states that anything that can go wrong ultimately will, makes no concessions for methods that merely decrease the odds slightly. And physical containment gives us no assurance that the escape that is bound to happen will occur later rather than sooner. Reasons like those outlined above make it clear why the politi-

cally minded advisory committee jumped at the
opportunity to demand a second kind of containment that,
when used in conjunction with physical containment,
seemed to eliminate the chance of accidental escapes.

Biological Containment and the Nasty Question of *E. coli*

Biological containment is based upon the principle that
any organism which requires the special, unique conditions
that a laboratory can provide will have a difficult time sur-
viving in the outside world, where few special favors are
ever granted. In the case of biological research, a not un-
reasonable corollary assumes that the more crucial the organism's needs, the less chance it has to survive.

The guidelines therefore designated three levels of biological containment within which recombinant experiments
could take place: EK1, for experiments acknowledged to
be safe, which require no special organisms (by using the
laboratory strain of *E. coli*, K12, the EK1 level of containment purports to provide "moderate" containment, since
experiments have shown the K12 strain to be far less apt
to survive than wild strains of *E. coli*); EK2, designating
host-vectors which have been genetically modified and are
predicted to provide a high level of biological containment (in which only one in 10^8—100 million—organisms
can escape); and EK3, designating EK2 host-vectors
which have been proved to be as safe as the predictions
assume and which will be used only for the most potentially dangerous recombinant experiments.

At the time that the guidelines were being drawn up,
there were no EK2 strains in existence. The workshops at
Asilomar had predicted with confidence that EK2 strains
could be developed within the year; yet it was almost two
years before Roy Curtiss III of the University of Alabama
actually constructed one. It now exists; but, since its efficacy has not yet been demonstrated by experimentation, it
still does not qualify as an EK3 organism.

The guidelines assumed that a combination of physical
and biological containment would provide the necessary
protection from whatever dangers recombinant experiments contained. Although risks still existed, the pro-

BIOLOGICAL CONTAINMENT (FOR E. COLI HOST SYSTEMS ONLY)

PHYSICAL CONTAINMENT
P2
P1

EK1	EK2	EK3
DNA from nonpathogenic prokaryotes that naturally exchange genes with E. coli Plasmid or bacteriophage DNA from host cells that naturally exchange genes with E. coli. (If plasmid or bacteriophage genome contains harmful genes or if DNA segment is less than 99 percent pure and characterized, higher levels of containment are required.)	DNA from embryonic or germ-line cells of cold-blooded vertebrates DNA from other cold-blooded animals and lower eukaryotes (except insects maintained in the laboratory for fewer than 10 generations) DNA from plants (except plants containing known pathogens or producing known toxins) DNA from low-risk pathogenic prokaryotes that naturally exchange genes with E. coli Organelle DNA from nonprimate eukaryotes. (For organelle DNA that is less than 99 percent pure, higher levels of containment are required.)	DNA from nonembryonic cold-blooded vertebrates DNA from moderate-risk pathogenic prokaryotes that naturally exchange genes with E. coli DNA from nonpathogenic prokaryotes that do not naturally exchange genes with E. coli DNA from plant viruses Organelle DNA from primates. (For organelle DNA that is less than 99 percent pure higher levels of containment are required.) Plasmid or bacteriophage DNA from host cells that do not naturally exchange genes with E. coli. (If there is a risk that recombinant will increase pathogenicity or ecological potential of host, higher levels of containment are required.)

PHYSICAL CONTAINMENT

P4	P3
	DNA from nonpathogenic prokaryotes that do not naturally exchange genes with *E. coli* DNA from plant viruses Plasmid or bacteriophage DNA from host cells that do not naturally exchange genes with *E. coli* (If there is a risk that recombinant will increase pathogenicity or ecological potential of host, higher levels of containment are required.)	DNA from embryonic primate-tissue or germ-line cells DNA from other mammalian cells DNA from birds DNA from embryonic, nonembryonic or germ-line vertebrate cells (if vertebrate produces a toxin) DNA from moderate-risk pathogenic prokaryotes that do not naturally exchange genes with *E. coli* DNA from animal viruses (if cloned DNA contains harmful genes)	DNA from nonembryonic primate tissue
DNA from nonembryonic primate tissue	DNA from nonembryonic primate tissue DNA from animal viruses (if cloned DNA does not contain harmful genes)	DNA from nonembryonic primate tissue DNA from animal viruses (if cloned DNA contains harmful genes)
"SHOTGUN" EXPERIMENTS USING *E. COLI* K-12 OR ITS DERIVATIVES AS THE HOST CELL AND PLASMIDS, BACTERIOPHAGES OR OTHER VIRUSES AS THE CLONING VECTORS	EXPERIMENTS IN WHICH PURE, CHARACTERIZED "FOREIGN" GENES CARRIED BY PLASMIDS, BACTERIOPHAGES OR OTHER VIRUSES ARE CLONED IN *E. COLI* K-12 OR ITS DERIVATIVES	EXPERIMENTS IN WHICH PURE, CHARACTERIZED "FOREIGN" GENES CARRIED BY PLASMIDS, BACTERIOPHAGES OR OTHER VIRUSES ARE CLONED IN *E. COLI* K-12 OR ITS DERIVATIVES

SOME EXAMPLES of the physical and biological containment requirements set forth in the NIH guidelines for research involving recombinant-DNA molecules, issued in June, 1976.

Reprinted with permission of *Scientific American* from "The Recombinant DNA Debate," by Clifford Grobstein, July 1977, vol. 23, no. 1, pp. 22-33.

ponents argued, the degree of containment demanded by the guidelines (see chart, pp. 146-47) gave more than reasonable assurances that novel organisms would not escape the lab or that, if they did, they would not survive long enough to cause harm.

Their thesis has been sorely challenged by opponents of the research. While few people have any doubt that the combination of physical and biological containment provides a measure of protection, not even the most ardent proponent can claim that these measures, by themselves, are certain to prevent an escape. Proponents maintain that the odds against an escape are so great that the actual infection of man by a bug carrying toxic genes is highly unlikely. Opponents respond by noting that highly unlikely is not enough. Their contention is that a slowdown in the progress of research could lead to further measures which could well provide almost absolute safety. And if the safety of the procedures is not improved, at least the researchers could turn to other hosts, ones not so compatible with man. Indeed, beyond the questions of containment and escape lies still one other bone of contention. That bone takes the form of a tiny, rodlike creature that scientists have always considered an accomplice in their research, a bacterium that has been used for almost fifty years in practically every biological experiment that has required the use of bacteria. It is a generally harmless bug with few toxic traits, but one as promiscuous with its genetic material as a wedding party is with rice. That bug, *E. coli*, has become one of the centerpieces of the debate.

E. coli and Man

Escherichia coli was isolated by Theodor Escherich, appropriately enough, from feces in 1885. Since then, its lot has done nothing but improve.

E. coli's normal habitat is the intestinal tract of humans and animals. In fact, its distribution in nature is directly related to the distribution of animal waste; because it is invariably found only in those areas that have been contaminated by feces, and because it is so hardy, *E. coli* is commonly used as an indicator organism which can detect the contamination of drinking and natural water supplies, as well as the presence of pathogenic organisms commonly found in feces in any number of environments. *E. coli's* ability to stay alive is also well known. It not only survives but multiplies in water; and while it is not known whether *E. coli* can multiply in soil, it has lived for up to fourteen weeks in limestone and up to twenty weeks in a pile of cow manure, without any outside assistance.

As an organism, *E. coli* displays the usual characteristics of a prokaryote: it has no nucleus; its chromosome complement consists of one or more loose threads of folded DNA; and it contains numerous plasmids, which are responsible for a variety of the organism's special functions, as well as for protection from external elements. It needs minimal nutrition to survive—sugar, nitrogen, certain minerals, and water—and has shown a marked ability to lengthen its survival time in climates with higher humidity.

Although *E. coli* is a tenacious little bug, it has its limits. Many, but not all, of the known strains of *E. coli* have shown their ability to colonize man's bowel. But artificially introduced strains usually survive for only a short time before the normal flora of the intestine drive them out of existence. In addition, strains that develop naturally in animals have a much harder time flourishing in the human bowel than the human strains, although some of them do survive.

Because of *E. coli's* strengths—its ability to survive, to multiply with very little care, to readily exchange portions of its genetic complement—scientists have long been enamored of it. And while it is true that the strain of *E. coli* that is used in most biological experiments—*E. coli* K12—is no longer strong enough to survive for any length of time in the human gut, it has maintained its propensity for untiring promiscuity in donating and receiving DNA, its ability to reproduce once every twenty minutes, its cheerful capacity to survive in practically any laboratory situation, and its relative lack of pathogenicity to common organisms. In short, it is almost an ideal bug for scientific experimentation.

Thus, by the time recombination became a viable tool for molecular biologists, *E. coli* had long been the organism of choice for almost any laboratory situation. Indeed, the simple fact of its past experimental history enhanced its usefulness as a host for recombined DNA. Present knowledge of the workings of *E. coli* is far more complete than for any other organism (between one-third and one-half of all its chemical reactions have been uncovered). This depth of knowledge has made *E. coli* a relatively predictable host; researchers can predict and confirm with great accuracy what each gene introduced into the *E. coli* host will do. That *E. coli* became the organism of choice in biological experimentation over fifty years ago was mainly a happy accident; that it continued to be used extensively was due to its own merits as an experimental organism; that it became the primary organism for hosting recombined DNA was a logical extension of its past. As James Watson noted, "Once serious work had started on *E. coli* . . . it obviously made no sense to switch to another organism if *E. coli* would be used."

There is, unfortunately, a momentum that can build up around almost any act or circumstance that makes it difficult to stop once it has begun. Reasons that seemed strong at the beginning might no longer apply, but the force of habit and the security of working with something familiar are enough to ensure broad support for a line of research or a course of action that should have been discarded.

So began the controversy surrounding *E. coli*. Researchers who wanted to maintain the rapid pace of research recognized the importance of using the organism. Changing to another would delay the progress of the research indefinitely, as research into the mechanisms of possible replacements was being conducted. Our knowledge of *E. coli* is one of the factors that makes it safe to use, they argued. Another organism, less well understood, might prove to have far more deleterious effects than one which has been studied extensively for fifty years. Besides, *E. coli* is a human symbiont; what great dangers could result from its use?

The opposition attacked these premises on two fronts. First, they maintained, only some strains of *E. coli* lived symbiotically with humans. Others were known to be pathogenic. Normal strains of *E. coli* can cause urinary tract infections—in fact, they cause the majority (70-80 percent) of them. *E. coli* has been known to cause gastroenteritis in babies, to be responsible for as many as one-third of all cases of meningitis in newborn infants, to be implicated in wound infections and up to 30 percent of all surgical infections. And it is now recognized that *E. coli* has been involved in causing pneumonia, although its role has not yet been defined. Therefore, while it is a symbiont, *E. coli* is also a pathogen. Genetic recombination might actually increase its pathogenicity, release other properties that we have not yet discovered, and transform this bacterium which is inextricably intertwined with our own lives into a formidable menace.

Second, *E. coli*'s ability to transfer genes naturally, while it is a great help to the conduct of recombinant experiments, also poses grave risks. Already it is known that *E. coli* can and does readily transfer genetic material both intra- and interspecies—both to other *E. coli* and to distinctly unrelated kinds of bacteria. The organism can effect this transfer in two ways: by transduction, a natural process which permits the exchange of genes both among bacteria of the same species and among those of different species (and which researchers have made use of to transfer genes artificially); and by conjugation, which permits transfer of long pieces of chromosomal and plasmid DNA within the bounds of a single species.

That *E. coli* has the capacity to transfer its genes to hosts other than its nearest relations is well documented. In the Philippines, for example, where prostitutes have been taking penicillin routinely for years as a prophylactic measure against the onset of gonorrhea, a new strain of the venereal disease has recently developed—one that has, as the cornerstone of its survival, the ability to resist penicillin.

The method by which the gonococci gained their resistance is enlightening. The preventive dose of penicillin habitually given to the prostitutes was smaller than the killing dose normally used to combat an existing infection. Ultimately, the gonococci learned to manufacture an enzyme, penicillinase, which counteracted the antibiotic, nullifying its killing power and allowing the tiny organisms to survive. Enzymes in different organisms are often quite different from each other, even when they perform similar functions; they may differ in weight, structure, complexity, or a number of other measurable factors. But researchers discovered that the penicillinase which makes the gonococci resistant to penicillin is remarkably similar to an enzyme produced by one other bacterium—*E. coli.*

There are very few hypotheses which can suggest rational reasons for such a remarkable occurrence. It is unlikely—almost beyond the remotest possibility—that the similarity between the two enzymes is a coincidence. Far more likely is a small scenario devised by the researchers: the constant prophylactic dosage of penicillin worked to create a vacuum in the normal flora of one prostitute's intestines by killing the weaker, more susceptible bacteria living there and increasing the opportunity of those resistant to its properties to survive. Over the course of several months, a colony of *E. coli,* containing the resistant enzyme penicillinase, began to flourish in the relatively noncompetitive climate created by the constant dose of antibiotics. During that period, a particularly hardy colony of gonococci entered upon the scene. They survived long enough to accept a genetic transfer of the plasmid through which *E. coli* had manufactured its protection. In the vacuum of the bowel, the bacteria multiplied, became entrenched members of the community, and began to emigrate to other available climes. So the disease spread from person to person, and grew to epidemic proportions in the Philippines.

The entire process of transfer and survival required a confluence of factors both unusual and, obviously, possible. A vacuum had to exist for the resistant *E. coli* to flourish in large enough numbers to come in contact with the gonococci; the transfer of genetic information had to occur successfully, so that the new host not only picked up its wonderful new toy, but was able to have it expressed in the production of penicillinase; and the newly resistant gonococci had to combat the myriad perils of the bowel long enough to survive and replicate and become a viable part of the resident flora.

But it happened. And, having demonstrably happened, it can easily happen again.

There is a small footnote to this story. *Neisseria gonorrhea*—the newly developed strain—has a distant cousin of its own, *Neisseria meningitidis,* which is responsible for spinal meningitis. *N. meningitidis* lives in the mucus lining of the throat, and at present is still sensitive to penicillin. But today's relatively unfettered sexual practices may change all that, for it is no longer out of the question that *N. gonorrhea* will find a way from its home in the genitals to the doorstep of *N. meningitidis*. It might then quite easily transfer its newly acquired resistance to penicillin to its cousin, and revive the terrible threat of an epidemic of spinal meningitis. Although the chances of this occurring are small, they are there. In fact, many epidemiologists are simply waiting for the day when their first case of penicillin-resistant spinal meningitis surfaces.

The case history of *N. gonorrhea* illustrates the promiscuous capacities of *E. coli*. But it also indicates something about the fate of entire biological systems as well. One point of resistance, one transfer, can lead to entirely unforeseeable results in what seems to be a completely unrelated sector of the environment. The case of *N. gonorrhea* shows clearly that the consequence of any action affecting any portion of an ecosystem has the potential to affect, indirectly, the entire chain. In this way, a hybrid characteristic inserted into an *E. coli* may very easily escape detection in its immediate environment, only to surface in a place where it was never expected, a place from which it is almost impossible to trace the characteristic's origins.

Genetic Sabotage

Biological containment of both hosts and vectors in recombinant DNA experiments was designed to end the speculative problems of escapes. By taking an already weak organism—the laboratory strain, *E. coli* K12—and making it totally dependent upon conditions supplied by the laboratory (and by doing the same to such vector systems as bacteriophage lambda), researchers hoped that they could mute the controversy surrounding the use of organisms that posed potential risks to humans, animals, and a variety of ecosystems. After all, if an organism simply could not escape, theoretical harangues as to the dangers it might pose if it did escape would become largely irrelevant.

Work on weakening *E. coli* began immediately after

Asilomar in several laboratories. It was expected that the effort would take only a few months at most, for what was needed was little more than a transfer of genetic characteristics into or out of *E. coli* K12 to make it impossible for the bug to survive without artificial intervention. But *E. coli* was smarter than anyone suspected. It had evolved a series of "fail-safe" survival systems of its own that permitted it to withstand incredible environmental stresses and pass on its DNA. To protect itself, *E. coli* can induce or repress literally dozens of enzymes in response to hundreds of environmental cues; it can grow and multiply, or simply stop growing, depending upon the conditions for survival and the availability of food; it can differentiate normally or form endospores—tough little cocoons which can protect it until crises pass; it can mate sexually or asexually, according to the dictates of its environment; and it can adapt to a lack of one of its nutrients—for example, the sugar lactose—by using, in its place, as many as a hundred other substrates. Its defenses, designed to ensure the survival of at least some members of each colony in almost any type of environmental condition, threw block after block in the path of the saboteurs. As one researcher noted, "*E. coli* has so many backup systems that it makes NASA look sick." (Researchers were having an equally tough time with the vector systems—bacteriophage lambda or plasmids like Co1 E1—because, as simple, metabolically inert organisms, they concentrate their entire energies on reproduction. Such simplicity yields few weaknesses.)

The battle lines were drawn: on one side were the researchers, working to disarm the organisms; on the other, *E. coli*, fighting just as hard to stay independently viable. But for the researchers, confrontation existed on more than one front. In addition to the resistance of the organisms themselves, they had to battle certain biological certainties that complicated their efforts. First, they found it extremely difficult to keep mutations that they transferred into the *E. coli* working through the generations. Artificial characteristics, while they could be successfully introduced into the bacteria, would express themselves for only a few hundred generations before they would disappear in the evolutionary process. After a while, it was found that deleting characteristics from *E. coli* produced far more stable changes than adding them. Second, while they could test their new strains in a multitude of environments, their lack of understanding of the ecology of an organism as ubiquitous as *E. coli* made it almost impossible to test the organism in all the situations to which it might be subjected. The number and type of environments in which a wild type of *E. coli* might survive is limited only by the composition of the sewage leaving almost every human waste facility. To test them all was impossible.

Nevertheless, Roy Curtiss III and ten of his students at the University of Alabama grew, isolated, characterized, and selected desired mutations until they came up with *E. coli* X-1776, a new, hothouse variety of *E. coli* K12 so sensitive to external factors in the environment that its chances of survival outside the laboratory are almost nonexistent.

Curtiss' bug was sabotaged in over a dozen ways. The new organism has the following characteristics:

- It requires diaminopimelic acid (an unusual amino acid not prevalent in nature, and necessary for cell-wall synthesis) and thymidine (for DNA synthesis) to survive.

- It has a generation time two to four times longer than wild strains of *E. coli*, which will prevent it from competing with the healthy, robust strains in nature.

- It is sensitive to bile salts, and cannot survive in the human intestine.

- It is sensitive to ultraviolet light.

- It is more sensitive to many antibiotics, drugs, and detergents than its parental strain.

- It cannot be productively infected with most known *E. coli* transducing phages under even ideal laboratory conditions.

- It is unable to inherit most conjugative plasmid types and is defective in its ability to inherit the rest.

- It cannot survive passage through the intestinal tracts of rats.

- It dies, degrades its DNA, and disintegrates outside its culture media.

- It dies relatively quickly after drying out or when it is suspended in water or physiological saline solution.

- It cannot transmit genetic information to other robust strains of *E. coli* by transduction or conjugation at detectable frequencies.

- It has a self-destruct capacity which kills it at any temperature below 42°C (human body temperature is 37°C).

Work has gone ahead on constructing safer vectors as well. Herbert Boyer and Charles Yanovsky of Stanford University and Donald Helsinki of the University of California in San Diego have been studying plasmids, hoping to provide the more common vehicles—like Col EI—with easy-to-read genetic markers that pose no threat to the usefulness of any of the more widely distributed antibiotics. They have also been attempting to endow Col EI with mutations that increase its sensitivity to changes in temperature and prevent its reproduction in any host other than *E. coli* X-1776. Others have been working with bacteriophages, concentrating on those which can combine with large pieces of DNA; a small group, called Charon phages, after the ferryman on the river Styx, has been found to kill any cells it enters so efficiently that only one organism in a trillion can survive its infection. Plasmids and viruses like these significantly increase the effectiveness of biological containment in many of the planned host-vector systems.

Despite the efforts that went into the development of these attenuated organisms, opponents of *E. coli* continued the fight. Even if the new host-vector systems created a reasonably effective barrier against escape, they still seemed inadequate; "reasonably effective" was, at best, meeting a minimal standard of containment, for the risks of recombination seemed in no way "reasonable" when compared to the risks associated with all previous basic research. Using a human symbiont, under any conditions, continued to gall opponents of the research. Other bacteria, like *B. subtilis* and pseudomonas, were available and amenable to research. And even though it would take from two to five years to bring the level of knowledge about these organisms up to that about *E. coli,* the delay seemed a small price to pay for security against possible infection by recombined bacteria. At the very least, use of these bacteria would preclude the kind of epidemic that might spring up within the population of a laboratory if an escape should occur. If that did happen, the barriers to research that were being constructed in the anticipation of potential risks would be minor irritants compared to those that would attend the ensuing public furor. Noting the distinct possibility of just such an occurrence, Robert Sinsheimer of Cal Tech remarked, "I can think of nothing that would impede science more than if in a couple of years there were an epidemic around Stanford or Cold Spring Harbor."

Biological and physical containment, therefore, have their problems in controlling those parameters of research that they are tailored to control: escapes of organisms from the confines of the laboratory. But other parameters, posing equal potential for escape, have been ignored. For example, containment is focused upon preventing movement of organisms out of experimental cultures; it does little to prevent the incursion of wild-type bacteria into the laboratory. Bacterial and viral contamination of laboratory cultures is one of the more common problems of microbiological laboratories. Despite the extreme precautions that may be taken to prevent escapes, it is quite possible that DNA implanted in *E. coli* X-1776 could find a willing host either in cultures contaminated by escaping *E. coli* X-1776 or in cultures of *E. coli* X-1776 contaminated by the outside world. Once biological containment has been overcome, the new host would have to bridge only one barrier—the weakest of all, the one most susceptible to human failings, physical containment—to escape into the world.

Whether or not escapes do occur in the future, that they

indeed can is evident from the very nature of containment procedures. Safeguards which must exist to make containment truly effective—mechanisms which detect escapes, for example, or those which cancel out the variables of human error—simply cannot be established in a laboratory studying recombined organisms. Thus, no matter how careful the scientists, no matter how precise the mechanical barriers, no matter how attenuated the experimental organisms, escapes from contained environments can and will occur, without the knowledge of those in charge.

Scientific Accident

Laboratory infections resulting from work with toxic organisms are a recognized part of the risks in scientific research. Although such infections are rare in comparison to the millions of scientific experiments that are performed each year, the dangers are an accepted fact of life among researchers. In fact, surveys of the medical and scientific literature have turned up over 900 incidents of infection involving more than 5,000 people and over 100 disease agents since such records have been kept. Not all of these infections were a result of human error; not all occurred as a result of the breaching of physical barriers of containment. But accidents resulting in infections have occurred often enough to make human error and the mechanical breaching of containment a real and significant threat to the safety of laboratory personnel.

Dramatic cases of accident and infection abound:

- Among disease agents, simian viruses—viruses that cause disease in monkeys—are among the most frequently studied, in part because of their similarity to agents of human disease. Scores of examples of the transmission of these agents to man exist. One, known as B virus, has produced twenty cases of infection; only three of the victims survived. Of the twenty cases, three were the result of laboratory accidents. One technician became infected when a cut on his hand was contaminated by saliva from a rhesus monkey. Another cut his hand on a broken bottle containing B virus culture. A third was preparing a monkey's skull for study.

- Marburg virus—another of the simian viruses—caused thirty-one infections in West Germany and Yugoslavia. Twenty-five of these infections occurred when laboratory personnel came in contact with the African green monkeys themselves, their blood, or their tissues. Two of the infections were the direct result of laboratory accidents. In one case, a physician pricked herself with a needle used to obtain blood from a patient. In the other, a technician carelessly shattered a test tube containing contaminated material. In addition, one case was discovered in the wife of a convalescent patient. Tests indicated that she became infected through sexual intercourse. Other cases of infection by Marburg virus occurred in personnel who performed autopsies on the dead monkeys, who handled discarded equipment, and who dealt directly with the infected animals themselves.

- In February 1976, two janitors at the Center for Disease Control in Atlanta, Georgia, contracted Rocky Mountain spotted fever and died. The subsequent investigation showed that they could only have caught the disease at CDC. Containment procedures for Rocky Mountain spotted fever lie somewhere between the P2 and P3 levels proposed for recombinant DNA research.

Other incidents of accidental infection in laboratories have been found throughout the scientific community. From the outbreak of smallpox that so sensitized Great Britain at the beginning of the debate over recombination to the epidemic of Lassa fever that claimed the lives of two researchers at Yale University and several missionaries in Nigeria, overt cases of accidents and infections have been documented through the years. Furthermore, the cases that have been published all share one thing in common: each one had to involve an infection serious enough to produce symptoms worthy of clinical attention. There is no record of the countless days of malaise, of the headaches or short bouts with diarrhea, of the "common colds," aches and pains, or times of irritability that might have accompanied other, subclinical infections. There is no record of the times that laboratory personnel might have felt slightly sub-par, a condition that in the rest of society is considered a normal response to moments of stress, or "something you ate." As a result, there is no record of the *real* incidence of laboratory infections, caused by any number of the thousands of possible human or mechanical errors that might occur within a laboratory setting. In a sense, infections like these make the prospect of recombination far more risky, for the more traumatic diseases that might escape a laboratory wear the red flags of flagrant illness. It is the little problems—those that can go undetected for weeks, months, or even years—that have the best chance of escaping and infecting the population in epidemic proportions. And they are as much a part of the risk of recombination as the disasters.

Accidents and the NIH Guidelines

Accidents in scientific research can have either positive or negative consequences. The problem with recombinant DNA research is that all the foreseeable accidents seem to be negative. And while it is true that accidents like the discovery of penicillin are hardly foreseeable, it is also true that they are extremely rare.

Still, one kind of accident envisaged by those who oppose recombinant research probably will not happen. That accident surrounds the mythology of the "brilliant high school senior."

The origins of this myth are buried by now in the murky depths of recombinant DNA's history. Most likely, it arose at an early press conference; when asked how easy recombination actually was, some scientist, searching for a metaphor, undoubtedly said, "It is so easy that a high school science major could perform it." Hyperbole became rumor; myth became fact; and experiments in recombination became objects for take-home chemistry sets.

The zenith of the myth arrived, finally, in the July 16, 1977, issue of the Boston *Real Paper*. Under the heading "Doing Recombinant DNA Experiments at Home: A Recipe for Botulism Soup" was printed an actual, but incomplete, procedure for recombining and growing a botulinus toxin. The steps and their explanations were outlined; the entire procedure was estimated to cost $350.

Yes, the time has finally come for us to realize that, in the comfortable surroundings of our own homes, we too can create new and different forms of life. But what we cannot do without the necessary equipment (which costs thousands of dollars) is examine our biological stew for signs of our success. Short of tasting the soup himself or serving it up to his younger brother or pet cat, the resident genius would have no way of knowing whether he has actually grown recombined DNA or has been left with a froth of healthy, normal, boring *E. coli.* Besides, if he really wanted to unleash a bit of botulinus toxin, he would need nothing more than an empty peanut-butter jar, about a quart of corn oil, a pound of fresh mushrooms, and a little patience. With these few ingredients, he could cook up a batch of botulism soup that would be the envy of any self-respecting molecular biologist.

But with some scientists turning their energies to devising ways around the guidelines, with others paying them no attention whatsoever, with the free-for-all atmosphere that exists in university graduate laboratories around the country, and with only those accepting funding subject to present regulation, the brilliant high school senior is the least of our worries. The guidelines themselves are shot full of holes, some of which practically invite the escapes of novel organisms.

From the time of the Berg letter in July 1974 until the guidelines were finalized in June 1976, there were multiple opportunities for a far more careful analysis of what the proposed guidelines might require. The situation was unique; scientists had agreed first to support a moratorium, lasting eight months, then to abide by the recommendations of the Asilomar Conference, which extended another fifteen. What the NIH could have done in this period of grace and what it actually did bear little relation to each other.

The NIH should have used the moratorium to finance directed preliminary research in several crucial forms. All proposed hosts—and especially *E. coli*—should have, once and for all, been tested for their ability to survive in the gut under most of the conceivable scenarios for contamination. This testing could have included an exploration of the different potentials for infection in laboratory personnel on regimens of antibiotics (a situation which could make them far more hospitable hosts for escaping organisms).

Methods for finding escaped organisms in the intestines of potentially infectable laboratory personnel should also have been explored. As it stands now, monitoring is an almost impossible task, for how does one monitor for escaping bugs that are basically similar to those we normally carry? And how can laboratories check for the transmission of genetic hybrids unless they have a clear picture of what those hybrids consisted of in the first place? If, however, the NIH had tendered a lucrative contract for laboratories working exclusively on developing genetic markers for the hosts of recombined DNA, perhaps some form of identification might now exist.

Working on strains like those needed to help ensure the safety of DNA work is thankless work, seldom recognized as ground-breaking, almost never in line for some of the prizes that make pioneering research so attractive. But it is work that is as crucial to scientific advancement as cooks and quartermasters are to the success of an army. Without the support of the kinds of clear, decisive research which determine the boundaries within which the flamboyant work is performed, the practice of science becomes little more than glorious skirmishes far beyond the battle lines, potentially dangerous exercises designed more to raise the morale of the populace—or the ease of cash flow, as the case may be—than to actually extend knowledge which might translate into social progress.

In addition, the system of supervision envisioned by the creators of the guidelines seems to have little basis in reality. In their view, it should be left up to "biohazard committees," which are essentially in-house organs of supervision that, in most cases, can be expected to practice little more than conventional brotherly love. Indeed, similar experiments in hospitals, in the form of morbidity and mortality committees, have been known to act almost as propaganda machines for the physicians brought in front of them. Designed to inquire into questionable deaths and implications of errors, these committees permit a doctor to state his point of view before they recognize the variables that can cause any medicinal procedure to go wrong and, like the lenient Catholic priest that everyone secretly wished for in confessional, let him go with minimal stigma.

It seems the height of insensitivity for the NIH to expect that such biohazard committees will be accorded the kind of respect and attention that their duties will require. Recombinant DNA research is not only a potentially risky field, but one that excites volatile emotions on both sides. Its supervision should be entrusted to neutral bodies, weaned of their professional affections, tied strongly to public participation, and acting in explicitly unbiased fashion. Indeed, one wag even suggested that, in the light of the fierce competition that has evolved in researchers' attempts to achieve that next significant breakthrough, they be permitted to investigate each other, and force each other to adhere to the guidelines. The result might be a level of conscientious supervision seldom achieved in the annals of civilization.

Federal laws are being prepared which may rectify the worst of the guidelines' loopholes. It has been proposed, for example, that the guidelines be extended to include in their scope industrial experiments in recombination, experiments over which the guidelines neither have nor imply any kind of control. It is also likely that some other form of scrutiny in addition to the biohazard committees—perhaps independent agencies with teeth to enforce their decisions—will be authorized. Nevertheless, if the guidelines submitted by the NIH are the best that the scientific community could come up with, questions as to its ability to police itself must be raised. True, it was the scientific community that exposed the problem in the first place. True, its actions at Asilomar formed the basis for the first guidelines which would oversee recombinant DNA research. True, it requested, on its own, the NIH's action in creating the advisory committee which drew up the final set of guidelines. But the results of even these conscientious actions show either extreme naivety or callousness; anyone who would have expected the guide-

lines—and, most important, their means of en-
forcement—to be the last word in regulation was sorely
mistaken.

The Mathematics of Risk in Recombinant Research

The risks in recombinant DNA research, when viewed
from the perspective of the procedures that have been
created to prevent them, are thus pretty grim. Containment does not work effectively; the guidelines do not supervise with any meaningful strength; accidents of all
kinds are bound to happen; and recombinant materials
quite simply will not be contained in the surroundings to
which they are supposed to be limited. But none of these
factors deals with the other partially quantifiable question
of DNA research: assuming that hybrid organisms can escape, what are the chances of infections and epidemics
breaking out?

One prominent scientist involved in performing the
research noted that opposition to it revolves around more
supposition than fact, more fantasy than reality: "Somebody comes [to me] and says . . . 'What you are doing is
dangerous because you put in this DNA and then *maybe*
the bacteria will grow [after it escapes], and, if it does
grow, *maybe* it will liberate DNA, and if it does liberate
DNA, *maybe* it will get into a cell, and if it does get into
a cell, *maybe* it will cause cancer, and if it does cause
cancer, *maybe* it will produce more viruses, and *maybe*
then these viruses will be infectious.' But nobody has yet
really found a mammalian tumor virus which spreads in
any sort of infectious fashion. So [the whole thing] is an
improbability built on implausibility."

What the scientist was pointing out was that there is no
direct relationship between escape and epidemic; a series
of mutually exclusive episodes must occur in a specific order before an escaping organism can cause even the remotest minor epidemic. The problem is analogous to the
chain of enzyme reactions that are necessary before the
genetic characteristic of skin color can be translated into
actual melanin and subsequent skin pigmentation. If the
chain breaks down at any one point, the translation of ge-

netic instructions to physical characteristics does not occur. The same is true of a potential escape by a novel,
recombined organism: if the series of necessary occur-
rences can be stopped anywhere during the process, epidemics simply will not happen.

Such a situation permits us to begin to quantify the
chances of accidents causing catastrophe during the performance of basic recombinant DNA research. By plotting
the precise steps necessary to achieve an event of catastrophic proportions, and by examining the odds of each
one happening, we can gain reasonable insight into the
probability of escape, infection, and epidemic.

In an article entitled "Should Genetic Engineers Be
Contained?" (published in the February 17, 1977, issue of
New Scientist), Dr. Robin Holliday did precisely that.
Taking an initial assumption of the worst possible scenario, he analyzed the scheme by which a shotgun experiment performed in a laboratory without special
containment techniques could release dangerous novel organisms into the biosphere, causing a massive epidemic of
cancer. In his construct, each necessary event is designated
P, each deleterious consequence C:

P1 A careless technician in the laboratory, while pipetting by
   mouth (in direct violation of the guidelines), accidentally
   swallows several million viable recombined bacterial hosts.
P2 The bacteria survive and proliferate in his gut (again, in
   violation of the guidelines, the bacteria used in this hypothetical experiment are *E. coli* K12, the regular laboratory
   strain, which is far stronger than the attenuated *E. coli*
   X-1776, far weaker than wild-type *E. coli*).
P3 The recombined plasmid the bacteria contain is transmitted
   to the more viable bacteria already flourishing in the flora
   of the gut.
P4 The transmitted plasmid contains a latent mammalian tumor
   virus which is induced to proliferate, perhaps because a
   repressive mammalian gene has been replaced by bacterial
   DNA during recombination.
P5 The plasmid containing the virus' DNA is present in many
   bacteria and, in at least some, is induced to form viable
   virus particles (such a step requires that the virus go
   through induction, growth, maturation, and release in its
   new hosts—a situation which has never been observed for
   mammalian viruses in bacteria, although it has been at-

tempted experimentally). The mammalian virus must be able to replicate its genome, produce its own coat proteins, and perform all the other necessary viral functions in a prokaryotic environment.

P6 The virus is pathogenic.

C1 The carrier dies from infection.

P7 The carrier is resistant and transmits the virus to others by contact or other channels of transmission.

P8 Some of the individuals to whom the virus is transmitted are more susceptible to it than the original carrier.

C2 The more susceptible carriers become ill or die.

P9 Only a portion of infected individuals show symptoms, but all are infective during the period of viral incubation.

C3 An epidemic is possible—one involving cancer, if the virus is oncogenic.

P10 The virus is reinserted into the chromosome complement of the human host (by transduction from *E. coli*) and induces a small, malignant infection.

C1* The individual dies—the epidemic spreads.

P11 The virus is both transmissable and oncogenic.

C4 Proliferation with no immediate outward symptoms would cause a widespread cancer epidemic years later, with no visible tie to the original accident in the laboratory.

The pathways of infection leading to each instance of dangerous consequences would look like this:

$$P1 \times P2 \times P3 \times P4 \times P5 \times P6 = C1$$

$$P1 \times P2 \times P3 \times P4 \times P5 \times P6 \times P7 \times P8 = C2$$

$$P1 \times P2 \times P3 \times P4 \times P5 \times P6 \times P7 \times P8 \times P9 = C3$$
and:

$$P1 \times P2 \times P3 \times P4 \times P5 \times P10 = C1*$$

$$P1 \times P2 \times P3 \times P4 \times P5 \times P10 \times P11 = C4$$

Dr. Holliday recognizes that assigning hypothetical values to each of the probabilities is slightly specious. Some of the steps require completely unpredictable circumstances for them to occur; others can be approximated with only slight variations. But, taken together, they provide one concrete picture of the capabilities of certain types of recombinants to infect our ecosystem:

P1 perhaps one chance in one hundred (probability 0.01), assuming that laboratories conducting recombinant research permit the grossest possible negligence (other possible accidents—far more likely—might more realistically be assigned such a high value).

P2 and P3 one chance in ten (0.1), based upon experiments by Ephraim S. Anderson of Great Britain, which demonstrated the viability of *E. coli* K12 in the human gut.

P4 one in 10,000 (probability—0.0001).

P5 as a conservative estimate, one in 1,000 (0.001), considering that this step has never been seen under experimental conditions.

P6, P7, P8, P9 are difficult to assess; if the virus is pathogenic, it is probably transmissible as well, but also easily recognized. The probabilities of P6, P8, P9 happening are low, of P7 quite high. Conservative estimates might give a high probability (perhaps one in ten, or 0.1) to all.

P10 and P11 an estimate of one in one hundred (0.01), although it has never been shown that any human cancer has been caused by a virus (natural and man-made chemicals are far more widely accepted as the principal culprits).

With these estimates, we can calculate the odds of deleterious consequences (C):

$$C1 = 10^{-2} \times 10^{-1} \times 10^{-4} \times 10^{-8} \times 10^{-1} = 10^{-11}$$

$$C2 = 10^{-11}(C1) \times 10^{-1} \times 10^{-1} \times = 10^{-13}$$

$$C3 = 10^{-13} \times 10^{-1} = 10^{-14}$$

$$C1^* = 10^{-2} \times 10^{-1} \times 10^{-4} \times 10^{-3} \times 10^{-2} = 10^{-12}$$

$$C4 = 10^{-10} \times 10^{-2} \times 10^{-2} = 10^{-14}$$

Thus, if ten scientists in each of one hundred laboratories carried out one hundred experiments per year, the least serious, most likely accident ($C1$) would occur on the average of once in a million years.

Dr. Holliday's calculations obviously cannot simply be accepted at face value, but must be examined in the spirit of good, clean fun that pervades any debate that is both important and largely speculative. Changing the nature of the recombined gene from that of an oncogenic virus to one causing subclinical infections, like mild diarrhea or small muscular aches, would both remove the first two possibly dangerous consequences (nobody would die from the infection) and significantly increase the chances of an epidemic (without the deaths of susceptible carriers, the infection has a greater opportunity to spread; if it is subclinical in nature, it will probably remain undiagnosed and unchecked). Furthermore, other possible scenarios—the infection of beneficial environmental prokaryotes by hybrid viruses released by the host bacteria, or the spread of disease to plants or animals, where it is far less likely to be detected—also increase the odds of actual hazards occurring.

Whatever their application in the real world of recombinant research might be, Dr. Holliday's numbers do make a point. Threading the narrow strait between the Scylla of emotions and the Charybdis of self-interest, they return at least a portion of the debate to a proper focus. And what they imply is probably of more importance to science, society, and the future of scientific research than all the arguments, all the charges and countercharges, all the stout defenses and dynamic attacks combined. For the numbers say that there are two broad categories of issues here, as distinct and separate as day and night. The first, important in the short run, is the issue of basic research and its hazards; the second, with its overwhelming implications for the future, is the effect of the present debate on the course of science's relationship to society. The numbers, while artificial, do not lie. And they say that recombination, in its most basic, experimental stages, is probably no more or less hazardous than a dozen other scientific, military, or technological pursuits. Indeed, far more potentially dangerous situations might have been (and might still be) created by the moon-exploration efforts, by ongoing work with viciously dangerous organisms like the viruses causing Lassa fever, Marburg fever, or a score more diseases, or by experiments that circumvent the problems of recombinant research by growing potentially dangerous pathogens like SV40 in tissue cultures and purifying them in centrifuges as living viruses, rather than by merely adding one of their genes to *E. coli* and growing a liter of hybrid hosts. The numbers say, yes, a risk does indeed exist, but the consensus is that there is no progress without risk, no civilization without risk, no life without risk. Other situations far more intolerable in their ability to cause suffering are a part of everyday life—the automobile, for example, kills upward of 50,000 people and injures 2 to 3 million others every year. But these situations are accepted for the benefits they bring. Even movements to make such risky enterprises safer falter under priorities which rate the pocketbook a more valuable commodity than an arm. (Although it is true that people never believe that "it" will happen to them. They assume that the choice is between their pocketbooks and somebody else's arm, which makes the decision somewhat easier.)

Basic research into recombination has, by now, weathered massive investigations, intense and extraordinary scrutiny, and a fistful of hysterical attacks, and has emerged with few scratches. Indeed, the most recent public statements have been by scientists like Roy Curtiss of the University of Alabama renouncing their former fears because the evidence which seemed so strong several years ago has lost its vitality, buried under the scores of recombinant experiments that have been performed without accident or dangerous consequences since the moratorium was lifted after Asilomar. Dr. Holliday's numbers may have an extremely limited scope in their specific application, but they do not lie.

But while the basic research itself may have suffered superficial wounds in the course of the debate, the credibility of the scientific community has taken some direct hits. The game of numbers which proved so beneficial in Dr. Holliday's arguments reveals some dangerous excesses and insensitivities when it is turned around and applied to the professionals on both sides of the debate. It is numbers which mark the length of time the NIH had to perform substantive research which might have obviated the need for much of the ongoing debate, and numbers which describe the tiny percentages of concerned citizens from the public who were included in the process of decision-making. And it is a decided rejection of numbers, of quantifiable, palpable, verifiable evidence, which led otherwise respectable and eminent scientists to characterize their opponents as "shits," "stagnating old men," "publicity seekers searching for a cause," and "a Tammany Hall politician, always asking "What's your price?" These public sallies could not even touch the bitterness that surfaced in private discussions, where scientists would almost search out fellow scientists or curious reporters and talk, "off the record," of the deeply sinister and self-interested motives of everyone in the opposition.

The implications of Dr. Holliday's numbers are quite clear. They undercut the arguments of the danger of accidents in recombinant research, and place them squarely in the laps of the protagonists themselves. But there is another side to the questions facing basic research, one far more philosophical in nature. It is not reached by numbers, and so has caused possibly more trouble among scientists than any other issue. The issue, raised by Cal Tech's Robert Sinsheimer, combines scientific and ethical problems, and, in a way, alters the nature of the entire debate. It is a question that relates specifically to recombinant DNA. But even if it is answered in its present, tightly defined form, it is a question that will crop up in the continuing interactions between science and society.

Crossing Nature's Genetic Barrier

One of the major problems with trying to define each and every risk of recombinant research is our lack of a predictive theory of evolution. Retrospectively, we have some idea of how man came about; but we cannot even begin to guess where nature will take us from here. Our ignorance leaves us in a very uncomfortable position as we attempt to examine the future of biological research. Events and situations that we would like to be able to foresee clearly appear only as shadows on the horizon—amorphous, indefinable, and slightly ominous. Our inability to predict the future of the world's evolutionary patterns with any accuracy has lent credence to a major criticism of the continuation of relatively unfettered recombinant research—the proposal that, by recombining prokaryotic and eukaryotic DNA, we might be courting danger by crossing evolutionary boundaries that nature meant to keep inviolable.

Nobody, on either side of the debate, has tried to define the risks of crossing such a barrier—indeed, people do not even agree that such a barrier actually exists. But it is the very lack of information on the issue, the very lack of agreement upon even the most basic sets of facts, that makes the contention so powerful. There is nothing that can be quantified (although people have tried), no numbers that can be presented to sway majority opinion in either direction. Because of its ambiguities, the problem of the genetic barrier has become one strong in symbolism; in a way, it is to the debate on recombination what recombination has become to scientific research as a whole. Neither can be proved to be a valid issue, yet both cause supreme discomfort in the ranks of the establishment precisely because they cannot be quantified. Both imply that logic cannot always be stamped in black and white and that the way to ease the burdens of dangerous knowledge is not always to search for more knowledge.

The barrier argument divides along the following lines. Critics of the research are not simply saying that crossing prokaryote with eukaryote will result in novel forms of life that are incompatible with the evolutionary flow of nature; rather, they are pointing out that evolution itself seems to have placed its own barriers in the way of this type of genetic exchange, and that it has probably done so for a reason. If scientific technology now destroys the balance that has long existed, without first understanding why it evolved in the first place, there exists a clear possibility that some as yet unknown tenet of natural law will be broken, with dire consequences. Sinsheimer puts it this way: "Somehow it is presumed that we know, *a priori,* that none of these recombined clones will be harmful to man or to our animals, or to our crops, or to other microbes. I don't know that, and worse, I don't know how anybody else does either." Breaking the barrier between single-celled and higher organisms may permit bacteria to become endowed with genetic elements that previously belonged exclusively to higher organisms. One possible consequence could be that viruses which now infect only bacteria would gain the ability to infect eukaryotic organisms as well. And against the newly infectious viruses, eukaryotes would have no natural defenses.

The responses to this argument are varied. Many scientists have pointed out that nature has provided so many opportunities for the exchange of genetic material that, undoubtedly, somewhere along the evolutionary path, many of the transformations being planned were carried out naturally and failed to survive the tests of viability that every new adaptation of an organism must undergo. The implication is that the barrier between prokaryotes and eukaryotes exists not at the point of genetic exchange, but at the level of the battle for survival. But to say that the artificial genetic exchanges being planned hold little danger because it is possible that similar exchanges have occurred in nature is begging the question. The question, in its simplest form, is whether we can rule out the crossing of a natural barrier as a possibly dangerous genetic event.

Sinsheimer's prestige and the relative calm of his criticism have served to make his arguments some of the strongest against the proliferation of recombinant DNA research. But the form of his questions and his status in the scientific community are only emotional calling cards; the real strength of his arguments lies in what they imply. And they imply volumes. They are phrased in a way that is conveniently general, leaving little that requires a specific defense, and nothing into which someone else could sink a fact-filled hook. They therefore attack the problems of research along an entirely different front than arguments dealing specifically with the risks of, for example, accidental escapes, for those arguments demand the imposition of facts (which scientists, whose lives are spent pursuing facts, are more than ready to supply). Sinsheimer's arguments search for the nonspecific, the vagaries of philosophy and ethics, the impact of moral viewpoints. To respond to them by claiming some level of scientific or technical expertise is almost impossible.

The difference between the two battlegrounds is subtle, but distinct. It is embodied in Sinsheimer's very words: ". . . and worse, I don't know how anybody else [knows] either." The implication is that scientific attempts to find technical solutions to experimental risks—in other words, to characterize the debate as purely technical in formulation of both issues and decisions—are not enough. By arguing in this way, Sinsheimer is opening the door to the other side of the recombinant DNA debate, the side which is less distinct, not as technical, but just as valid. It is through this door that the questions of bioethics, and of science's relationship to society, will pour.

Sinsheimer's arguments also indirectly deride the characteristic scientific arrogance—the monolithic belief that everything boils down to solutions couched in physical and chemical reactions. Although this arrogance has practically overwhelmed the recombinant debate at times, it does not rest only with scientists. Professors of history have been known to announce only half jokingly that all problems are those of diplomatic history; some sociologists seem to feel that the answers to global problems revolve around the nature of the corner bar; and there are psychiatrists who like to trace nine out of every ten impulses to an affectionate mother or a distant father. But the arrogance in the recombinant DNA debate emanates from the scientists—and most of it comes from the group that is pressing for relative freedom for the research. Its attitude is not something that we can accept with grudging amusement; it is not the kind of egocentrism that permits us to walk away shaking our heads because we think that it cannot affect us. To the contrary, the arrogance displayed by scientists on this issue is particularly galling, precisely because it subverts discussion of an issue which strikes at the very heart of the nature of public debate and responsibility. The scientist is saying, in effect, "Trust me. You cannot be expected to understand the true complexity of the problem." And of his opponents: "They are blinded by jealousy. They have their own axes to grind. They refuse to see things which, as good scientists, they should recognize." Sinsheimer's response raises the question of who is saddled by the most self-interest. For, when push comes to shove, the only entity whose self-interest really matters is the one most affected, the public, who, as Roger Dworkin noted at Asilomar, has the right, through its legislative representatives, to make its own mistakes.

Through Binoculars, Backward

While it is evident that neither the scientific nor the public community has reached consensus on whether or not research should be continued at its present rate, it is obvious that—at least until further developments—researchers will be permitted to continue their efforts in some form or another. A second moratorium on the research is out of the question, barring any unforeseen catastrophe in the near future. And it seems as if the conduct of research for the next twenty or thirty years will be used as a yardstick to measure whether or not we have done the right thing in permitting it to continue. If an accident occurs, the shouts of self-righteous glee (tinged, perhaps, with a touch of sympathy for the victims) will no doubt rise from the ranks of those who have worked so hard to oppose unrestricted research. If nothing untoward happens, the researchers will go merrily on their way, concocting newer, ever more obscure organisms. And the basic research will continue to make inroads on the secrets of nature, leading us to the brink of a technology so fundamentally different from the ones we now know as to portend a dramatic revolution in the way we conduct our lives.

What strikes the hardest about such a prediction is how little it seems we have been able to discuss the "other" issues involved in the research, the issues which have been labeled irrelevant by the scientific establishment. We have, in a way, been cleverly deluded into looking at the object of our investigation as if through a pair of binoculars turned backward. A potentially great problem has been made small; circumstances which strike at the very heart of the way we live have been discussed simplistically by decision-makers, who have managed to define issues as if the only things at stake were technical in nature, as if ethics and morals were not involved.

Those who have been making the decisions indeed have had all the answers, but they have come up with pitifully few solutions. Compared to what has been left undebated, the questions that have been decided so far have been minor. The two-edged sword has, it is true, tested science's ability to maintain control over itself. And in that respect, the scientific establishment has handled itself well. It has recognized a problem, convened its meetings, discerned a valid solution, and (it hopes) moved on. But the scientists' definition of the problem in the first place and our own are two entirely different things. The first responds to data;

the second introduces variables that cannot be decided by facts, that will remain ambiguous, almost theological quantities.

It is quite obvious why scientists wanted to steer clear of the dangers of ethical arguments. Ethics are no longer the companions of scientific thought. Scientists are working to learn more and more about less and less. Many of them have learned to fear the uncertainties and ambiguities that go hand in hand with the greater issues that beset us, for those issues are always open to interpretation and never seem to get solved. For those scientists, responding to questions about scientific accidents, the efficacy of containment, and the relative dangers of a certain line of research is simple. The figures of one side are compared to those of the other. Numbers replace reality as surely as plastic infantry units stand in for flesh-and-blood soldiers in the Pentagon's War Room. And if the air above their battlegrounds becomes so fogged with detail that the larger, riskier issues are lost in the haze, so much the better. It seems far safer that way; by risking little, they lose little. By fighting carefully restricted battles, they cannot lose a war. Unfortunately, they cannot win one either.

So it is that both sides have been using the legitimate issue of recombination for larger purposes. The researchers, by concentrating on technical aspects (a fight they feel confident of winning), hope to divert attention from questions that could present serious problems to their freedom of inquiry. The critics are trying to force open that very issue—to initiate a confrontation on science's role in society—by using recombination as a bludgeon, whether or not it is a valid point of contention. Ulterior motive has turned a problem with serious implications for the public health into a political football. And much of its power has been dissipated in the raised voices and hyperbole that have turned even its most technical aspects into matters of principle.

Nevertheless, the larger issues will be heard. And their unveiling will have come about as a direct result of the issue of recombination. It is a small paradox that scientists fashion the very sword that hangs above their heads, for each purely scientific discovery that they make both expands the borders of their knowledge and impinges upon the right of society to decide for itself how it wants to live. In the future, it is this issue which will form the cutting edge of the debate, this question upon which the course of scientific discovery will ride. And whether the sword belongs to Damocles or to Alexander, cutting the Gordian knot is a decision that the scientists themselves have to decide. For it is now, before the dangers of unfettered research become almost uncontrollable, that decisions have to be made. When technology has outstripped our ability to cope and adapt, it will be much too late.

FIVE:
Research and Society

In 1831, Michael Faraday was discussing his revolutionary research into the powers of electricity when one skeptic popped up and demanded to know what good his discoveries really were. "Sir," responded Faraday, "what good is a newborn baby?"

Two babies have thus far surfaced in our discussion. The first, quite newborn, is the technology of recombination. It has lent itself to speculations encompassing both hope and doubt; its inheritance and environment have given us some clues as to how it will grow and develop. But many of its enigmas have been left unsolved. Part of the reason is that certain questions will, by definition, exist until the infant has reached a stage of maturity when it can respond on its own. But some of our questions can be answered if we turn our attention away from a direct examination of recombination and puzzle through its heritage a little more thoroughly. And its heritage lies in that second baby, science.

Although people have been inventive since the first semi-erect humanoid thought of using a stick to knock an unreachable banana out of a tree, science has not existed as a separate entity until quite recently. Indeed, until less than 300 years ago, science was considered another branch of philosophy; it gained and lost importance as its sister disciplines waxed and waned. It was not until the Industrial Revolution that science began to take on a character of its own—not until the early nineteenth century that the term "scientist" was even coined. At that point, society's demands for improvements in the way people lived began to catch up with the snail's pace of progress. People began to realize that science was a valid way of pushing forward; it was then that money and attention began to pour into research. Samuel Butler's observation that "all progress is based upon a universal desire on the part of every organism to live beyond its income" was founded on the unerring assumption that the first thing that almost every individual provides for is his or her material well-being; spiritual and emotional needs become important only when the belly is full and the feet are warm and dry.

Before the nineteenth century, public support of science was based strongly on altruism. The public funded symphonies, museums, universities; in the same vein, it supported free research, not necessarily because it was useful, but because it was part of the tremendous spectrum of intellectual activities in which man engaged. As merely one of so many, science received no more attention than any other intellectual endeavor. Indeed, its place and importance were relatively minor, for even major basic discoveries seldom led to immediate practical benefits: 150 years passed between the discovery of lighting by gas and its first semicommercial use; 200 years elapsed between the evolution of the principle of telegraphy and the first actual model; and between Avery and Berg—between the initial experimental recognition of the genetic significance of DNA and the first genuine attempts to direct its engineering—thirty years went by.

As the gap between discovery and use shrank, however, the value of science became clearer. From merely one of many, it slowly became separate and distinct. Like cream rising to the top, science began to surface as one intellectual pursuit that was going to pay cold, hard dividends.

In the United States, the rise of science coincided with the decline of the evangelical movement in the 1840s and the increasing importance of materialism. As "more" and "better" became the passwords of progress, science became known as the repository for many of their successes. Soon, it began to stand as a symbol for progress itself, with its promises of longer life, higher standards of living, more productive industry and agriculture, and the possible ultimate explanation of the existence and nature of the universe.

As science grew into the institutionalized version of truth and order, its practitioners, the scientists, became their priests. There was little, it seemed, that science would not be able to do, few diseases that it would not be able to cure, fewer machines and tools that it would not be able to invent. Furthermore, little of what science was doing could be seen to have any seriously harmful implications. While the practical work of turning scientific discovery into technology was being cranked into high gear, the philosophy of science was widely understood to be benevolent. Research stood out as an attempt to learn the truth about nature without necessarily making use of it or, in the end, harming it.

Scientists benefited from the warm glow of approval. Indeed, there was little reason why they should not, for they lived generally in society's mainstream. Few professional researchers even existed; to most scientists, research was a hobby. Professors of science made their livings teaching students, who, in their turn, would become professors themselves, unless they chose to work in such "practical" fields as engineering and industry. Society therefore perceived science and scientists as an inseparable part of itself, as its appendages, dedicated to extending the frontiers of truth and order a little further.

Such a view of science was based upon practical experience. That experience went down the drain with the unfolding of the first World War. Practically overnight, science metamorphosed from a discipline that could do little wrong to one that could devise the technology for bombs, warplanes, and mustard gas. War no longer involved merely soldiers; it enveloped populations. And the reason for the shift could be traced at least in part to the increasing sophistication of applied science.

With the end of the first World War came the recogni-

tion that what science had done in such a short time for
the various war machines it could also do for industry and
technology. It was only a small leap from transporting
bombs to transporting passengers, although the problems
of delivery were distinctly dissimilar. And the networks of
communications which had proved so valuable in a war
that stretched over fronts hundreds of miles long needed
little modification to make them suitable for commercial
use. So scientists turned to the hard science, the technological science, the science of quick, concrete results. As they
did so, they left a little of their earlier character behind.
From a group which thought largely holistically, seeing
science in terms of its relationships to larger religious, ethical, and moral problems, the scientists became reduction-
ists, concerned with learning everything there was to know
about one small segment of science, specializing within
specialties.

The reasons for this shift are understandable.
Knowledge was growing at an exponential rate; already
there was too much for any individual to handle. And developing a scientific philosophy about one's work—a task
which demanded inquiry beyond one's chosen field—
would only have taken time from the subject at hand. So,
faced with a choice between doing one thing well and two
things competently, most scientists chose to specialize.
Specialization demanded greater concentration on research;
increased concentration reduced each researcher's capacity
to deal with the philosophical side of his work and made it
almost impossible for him to concentrate on his students.
Many of those who worked the hardest—and who specialized the most—moved farther and farther from the main
current of society into the wilderness of pure research. In a
very short time, much of science became autonomous from
society, moving out of the public's orbit of understanding
and into fields where laymen could only hope that the
results of research coincided with their own desires.

Then, almost on cue, came an event of such magnitude
that it acted both as a catalyst for the continuing reduc-
tionism within science and as a warning signal for society
at large. The Second World War was, in a sense, the first
war that used science as its acknowledged principal
weapon. Population centers were no longer primary strate-

gic prizes of the opposing war machines—they were attacked more for the psychological and economic burden
the tacticians thought their destruction would bring than
for their military value. Instead, the great factories, the
war-weapons industry, and the research facilities struggling
to create such monstrous spawns of basic science as the
atomic bomb became the focal points for destruction.
There is considerable disagreement in historical circles
about the crucial turning points in the Second World War,
the moments that guaranteed an Allied victory. One candidate nominated time and again is the Allied bombing of
the German factories involved in manufacturing the heavy
water and radioactive materials needed for an atomic
bomb; for although disabling these factories might not
have ensured victory, it quite likely eliminated the possibility of a decisive defeat.

By the end of the war, the trickle of scientists involved
purely in research had become a flood. The knowledge
needed to do anything original had become so complex, so
massive, that researchers were forced ever deeper within
their specialties. As a result, they lost contact with some of
the great issues that had catalyzed their work when it had
taken place within the larger context of science and society. As Van Rensselaer Potter noted in his book *Bioethics:
Bridge to the Future*: "[Cosmic issues] were neither important, useful, nor interesting. It was assumed that all
new knowledge was basically good and that it contributed
to order in a way that, if not immediately obvious, would
become obvious in the fullness of time. If there were no
individual scientists able or willing to cope with cosmic issues, neither were there philosophers of the Descartes variety who could hope to comprehend all available
knowledge."[1] But while all knowledge might have seemed
good and constructive to many scientists, laymen were beginning to wake up to the possibility of dangerous
knowledge, of new, runaway truths, utterly beyond civilization's control:

The feeling grows that scientists are finding it increasingly difficult to predict the consequences of their work, that technology

1 Van Rensselaer Potter, *Bioethics: Bridge to the Future* (Englewood Cliffs, N.J.: Prentice-Hall, 1971), p. 58.

has become the sorcerer's apprentice of our age. The concept of dangerous knowledge appears in a variety of images—the mushroom cloud, the usurping robot, the armless child of thalidomide. Many scientists object violently to the idea of dangerous knowledge, taking the position that all increases in knowledge are inherently good. This attitude is undoubtedly interwoven with our religious heritage, which assumes that the world exists for the benefit of man and that human suffering and evil serve part of a greater purpose. [But] . . . the concept of dangerous knowledge is valid, if for no other reason than that it calls attention to one of the dilemmas of our society. Dangerous knowledge has been defined as knowledge that has accumulated faster than has the wisdom to manage it; in other words, knowledge that has produced a temporary imbalance by outpacing other branches of knowledge.[2]

Dangerous knowledge is that, and more. For Potter implies that the only way to protect ourselves against such dangers is to seek more knowledge—to seek new truths so that we can cope with the old. It is not unlikely that these new truths will contain their own elements of danger, their own risks, which can push us forward ever faster until we find ourselves in a never-ending spiral of action and reaction, of combatting imminent danger with future danger, of mounting complexity upon complexity, tool upon tool—until we finally have reached the limits of our ability to adapt. Man may now be at the pinnacle of evolution. But as Teilhard de Chardin noted, "There are no summits without abysses"; there is little chance that we will remain dominant unless we examine carefully our own future in light of the burdens and delights placed upon it by that wild-eyed adolescent, science.

But that is getting ahead of the chronology. In actuality, while the Second World War nurtured the growing unease that certain small sectors of society felt with the increasing impact of science, few among those in power paid much attention. Science and technology were the architects of our future society. Whatever had to be done to encourage them seemed reasonable. And there was little question that scientific research—having outstripped the tiny laboratories and glass labyrinths that symbolized the research of the 1920s—needed compensation for the "more" and "bet-

[2] Potter, p. 76.

ter" that it was giving to society. Simply put, it needed money, and lots of it.

Because the universities housing research could hardly afford to support the enormous requirements of modern science, the federal government stepped in on a massive scale. In biomedical research alone, the government's involvement shot up almost geometrically, from a low of $4–7 million in 1945 to $1.6 billion in 1968, about a 300-fold increase. And money for biomedical research was only a fraction of the government's contribution, which included sufficient funds to put a man on the moon after only twelve years of intensive efforts and enough money to give birth to an entirely new generation of military weaponry. Both these examples of applied research stemmed from the government's strategic and military requirements: the first came as a response to the Cold War, military competition, and the successful efforts of the Soviet Union to launch the first satellite, while the second came as a direct result of the war in Vietnam. Neither can be interpreted in any way as governmental sponsorship of free inquiry; both indicate that society's priorities supersede scientific prerogatives. Because the money was rolling in, researchers paid little attention to the fundamental change in policy that this kind of subsidy implied. Their metamorphosis to specialists was now complete; as long as the money and freedom to pursue avenues of research closely related to their own existed, it mattered little how the results were used. As long as they could continue their work, they could ignore the darker implications of massive federal funding of technologically oriented research.

After 1968, government funding of biomedical research leveled off. Inflation, the war in Vietnam, and the diverting of funds to Medicare and Medicaid saw to that. But the government has continued to be the primary source of funding for almost all forms of scientific research. As the principal sugar daddy of research, the government has also been in a position to enforce acceptance of its priorities, to act as the main arbiter of what its cash would and would not buy. Thus, its priorities have often been adopted by the researchers who have benefited from its largesse. Long-held loyalties have been shattered as the researchers turn from their traditional benefactors, the universities, to the government for support.

Universities, too, succumbed to the same problems, for they became dependent upon the overflow cash provided by federally funded research to pay for educational departments which otherwise operated at a loss. In this way, the microbiologist who received a grant from the government was indirectly supporting the professor of Chinese or art history as well as his own project. To adapt to the federal takeover of funding and the competition that it engendered, researchers focused on projects which seemed to fit federal demands most closely. The issue of freedom of inquiry, soon to be raised as a battle flag over the debate surrounding recombination, was dead practically before it was born.

With the advent of atomic weaponry, industrial pollution, biological warfare, oral contraceptives, and, finally, recombinant DNA research, science had come full circle. The explosive revelations produced by its work finally brought it back to direct contact with the public. Sooner or later the paradox was bound to surface. Society could not simply continue to hand over the key to its treasury without the assurance that the things it was buying could not harm it at some future time. The point of collision, when the risks of progress seemed to match or exceed its benefits, finally arrived with recombinant DNA research.

The relationship between science and society is now a very brittle one. Each has something the other needs, as well as something the other fears. Each is demanding its own set of rights—rights which, in their uncompromised form, are obviously incompatible with each other. While acknowledging that absolute rights do not exist (even freedom of speech does not extend to shouting "Fire!" in a crowded theater), scientists are demanding the right to continue their experiments along legitimate lines of research, wherever they may lead. The public, on the other hand, recognizes its obligation to protect scientific research, but must balance that against another fundamental obligation, that of protecting the welfare of its citizens.

Somewhere between these two absolutes a compromise does exist. But it must be one that is flexible enough both to protect the public from dangerous, unwarranted science and to ensure that scientists are not harassed unreasonably by an overly sensitive public. Such a compromise will be at least partly based upon the balance of hazard and benefit that accompanies a controversial piece of research. But, because hazards and benefits are both speculative and subject to interpretation (what one individual might view as an unconscionable danger might seem to another to be an acceptable risk), other criteria for judgment must also be developed. In the long run, it will be incumbent upon us to construct a fundamental moral philosophy around the concept of scientific research, one that defines our goals, our aspirations, and the limits to which we are willing to go. Such a philosophy cannot be thought of as an absolute, objective set of criteria, of course; its priorities will depend largely upon the mood of society and the state of its relations with science. But it can exist as a foundation upon which pragmatic decisions of policy can be made.

Dr. Daniel Callahan, the director of the Hastings Center (Institute of Society, Ethics, and Life Sciences), has devised three sets of these so-called "moral policy options" which seem to include most of the philosophical contradictions that exist in the relationship between science and society:

A. *Intervention or non-intervention into nature.* One can take the general position that mankind should not intervene into nature at all unless it is absolutely necessary for the sake of the preservation of human nature to do so. Alternatively, one could take the general position that mankind should feel perfectly free to intervene into nature as long as it suits his purposes and meets other moral tests. Nature, that is, can be construed as either sacred or . . . wise or both, and intervention approached with great wariness. For instance, it might be contended under such a policy that long-established evolutionary patterns in nature should not be tampered with. Alternatively, it might be argued that nature is neither sacred nor wise, and that intervention is perfectly appropriate if it promises to produce no practical harms.

B. *Risk-taking versus caution.* Here again one encounters a basic choice: one can decide that, all other things being equal, it is part of the glory of human nature (and especially of science) that it is willing to take risks and willing to gamble in order to understand life better or to improve life. Scientific progress is both possible and good, and progress is always to be preferred over the status quo, even if it is necessary to take some risk in the name of progress. Alternatively, one could take the position that caution and care is the best policy, that, all things being equal, it is best to adopt a stance of wariness, and to take only slow, measured steps toward change and innovation.

C. *Doing good versus avoiding harm.* Again, the fundamental kind of moral choice: ought the goal of the moral life, taken at least in the social and community sense, be that of seeking always to do good—to make life better, to cure illness, do away with misery, make more people happy; alternatively, is our moral obligation limited to the avoidance of doing harm to others?[3]

Outside of the considerations of time, these three criteria stand up well. If, for example, an individual is a gambler who believes that we have the right to intervene in nature and the obligation to do good, his stand on scientific research like recombination will be decidedly different from that of a cautious individual, whose priorities include the avoidance of harm and nonintervention into nature. Other combinations of the three lead to a variety of approaches to research, some of which are more active, some of which are more passive.

When each criterion is applied to a specific kind of research, the possibilities of this approach become clear. Let us assume, for a moment, that recombination, with all its attendant risks, is generally acknowledged to contain the secret—and therefore the cure—of cancer. Many otherwise neutral people might easily become gamblers, recognizing the existence of cancer as one of the principal banes of our lives. The problems of intervention into nature and avoiding harm might not seem so large, mainly because the opportunity for performing some substantial good is so readily available. On the other hand, if we can count on only a substitute for insulin as the principal benefit of the research, perhaps far fewer people would be willing to gamble with the risks to nature and the public

---

[3] Daniel Callahan, "Ethical Prerequisites for Examining Biological Research: The Case of Recombinant DNA," unpublished paper.

well-being, knowing that far fewer people would be affected, far less good would result.

There is a great difference between judging recombination through a pure assessment of its benefits and risks and recognizing the issue as one which we can readily look at through the clarifying prisms afforded by basic moral policies. The controversy over DNA by itself teaches us little, whether or not we resolve it. The next controversy, and those that follow, will each have to be decided upon its own merits. What has applied to one case will have little effect upon the resolution of others, and each situation will have to emerge from chaotic conditions before it nears any kind of reasonable compromise. If, however, protagonists can begin from a position of basic, widely recognized understanding, they can deal from accepted knowledge, whether or not they agree upon the speculative issues at hand.

What, then, of recombination?

It may already be too late to argue from a position of abstract moral perspective. Recombinant DNA research is and has been a powerful issue; devising guidelines under which the debate should continue will have little effect upon the ways it has already progressed. With the mood of today, it would be almost impossible to take the relationship of science and society out of context, work out a reasonable set of guidelines, and return to apply them to recombination without being vulnerable to the charge that it was the issue that determined the guidelines rather than the other way around. Furthermore, the issue itself is so distorted by emotions on both sides that we are simply going to have to muddle through, treating it both as a legitimate problem in itself and as a teaching tool for the future. It is important that the issue of recombinant DNA research be settled in a way that protects both the public and the relative security of scientific research; but it is just as important that this settlement be reached in a way that does not preclude fair and rational discussion of future issues. Recombination is the frontier, the first outpost in previously unknown, unrecognized terrain. If it is explored carefully and well, we can pave the way for a rational handling of other problems; if it is roughly exploited, the relationship between science and society may be compromised for decades to come.

What Society Owes Science

In the 1960s, studies of the genetic composition of male inmates in "mental-penal" institutions revealed that many had abnormal sex chromosomes. Instead of the usual double—XY—complement, approximately 2 percent of these inmates were found to have a triple—XYY— genotype. Since the incidence of the XYY genotype throughout the general population is roughly 0.11 percent, the significant increase of XYY in penal institutions led many to hypothesize that the abnormality was an indicant of criminal behavior. While it was true that the vast majority of people with XYY chromosomes did not end up in institutions, the statistical frequency of the genotype generated a variety of studies which examined the behavior of newborn infants in an effort to compare the progress of the affected children with that of children with normal chromosomal complements.

One of these studies, conducted in Boston by Dr. Stanley Walzer and Dr. Park Gerald of Harvard, consisted of a screening of all newborn males for both XYY and XXY defects, as well as a follow-up study of the psychology and behavior of the children afflicted. The study began quietly enough in 1968; but in 1973, it aroused the interest of a group of public advocates, Science for the People (SESPA). What ensued constituted a war, a brutal, bloody battle over scientific and emotional issues surrounding human experimentation.

Science for the People, and other concerned citizens as well, objected to a variety of assumptions in the Walzer study. The group criticized both the theory and the conduct of the research, claiming that the results of the study were rendered meaningless by parental knowledge of the child's abnormality, since it could not help but color the way the affected child would be viewed and treated as he grew. In addition, the very assumption that the XYY genotype implied criminality was a societal prejudice based upon an ideology of eugenics "directed against the poor, minority groups and supposed social deviants." Dr.

Walzer's study, claimed the group, was biased before it ever began by the initial crippling publicity given the XYY genotype; it was rendered useless by its own methodology, which permitted both investigator and parent to know which children were affected, a situation which could only lead to the self-fulfilling prophecy that XYY was indeed an antisocial characteristic; and it was harmful, in that it took a child who might otherwise have grown up normally and subjected him to the risks of intense outside scrutiny and altered parental responses.

The group's charges were serious. Dr. Walzer's study was reviewed by the Harvard Human Studies Committee, which is required to evaluate all grant proposals to determine whether they comply with rules about informed consent of human subjects. Both sides aired their opinions, in front of the committee and in the press. After hearing all the arguments, the committee decided to permit Dr. Walzer to continue his research, agreeing that there was a need for new information on the XYY genotype and that his study could provide it.

Then the war began. Dr. Walzer began receiving telephone calls, threatening him and his family with bodily harm. In the meantime, Science for the People claimed that a "reign of terror" had been instituted by the university establishment, and that a number of nontenured Harvard faculty members on the Human Studies Committee, while sympathetic to the SESPA point of view, "could not vote for us in public for fear of their jobs." In genetic research, as in love and war, all seemed fair.

In the end, Dr. Walzer voluntarily suspended his research despite the vote of confidence he had received. But the end of the project could not wipe away the ugly picture of wildly hostile antagonists shucking their responsibilities to their constituencies like dirty clothing and knuckling it out in the mud. Valid arguments on both sides had been buried beneath the emotionalism and self-interest that characterized both the defenders and the critics. And while little research is now being undertaken into the significance of the XYY genotype, most people agree that the true issues were never really resolved.

In trying to define society's obligations to science, the first question that must be asked is who, in fact, consti-

tutes society—or the public, since we have been using the
terms interchangeably. Although characterizing society in
general is easy, the constituency interested in any single
issue is far smaller, and is limited to those who demonstrate an active involvement in finding solutions. In trying
to determine who "the public" in the recombinant DNA
issue is, we are not discussing those affected by the problem but those who are trying to influence its resolution;
the public involved is as separate from society as a whole
as a cadre of farmers disturbed by a wheat deal with the
Soviet Union, or a civil-rights group lobbying for busing to
achieve racial integration. While each of these issues affects all of society in some way, only those showing active
involvement—only those who become advocates for the
public welfare—can pragmatically be termed "the public."
By virtue of their involvement, they become our representatives.

In the case of recombinant DNA research, the public
consists mainly of state and local governments or their
designated committees (in Cambridge, Massachusetts, for
example, and in California), various public interest organizations—Science for the People, Friends of the Earth,
the Sierra Club—and, on a national level, Congress and its
committees. Members of these bodies have demonstrated
particular interest in the side of the question that attaches
priorities to the public welfare and have tried to become
actively involved in the making of decisions.

In general, society's single most important obligation to
science is to remain passive. For ethical and practical reasons, its duty is to leave basic scientific research alone as
long as the public welfare is not threatened. Furthermore,
its role in regulating or overseeing science when necessary
must be both reasonable and restrained; oppressive societal
pressure upon scientific research—as in the case of the
threatening telephone calls during the XYY controversy—
can only lead to intransigence by researchers. To put it
more simply, society has the fundamental right to protect
its own welfare. But that right, while theoretically absolute, must give way to the practical recognition that it is
the researchers themselves who can best understand and
interpret the technology of their work. Our right to regulate is thus contingent upon their willingness to share their

knowledge. Turning the screws to science may get us out of the recombinant DNA issue by foreclosing it completely; but to do so would practically ensure future non-participation by scientists in issues of similar importance.

While such a position might sound like a craven compromise, it is not. Society's involvement in the procedures of basic research is, by definition, a negative involvement: public advocates do not set their sights on a piece of research to encourage its continuation, but to inhibit or stop it. What we must try to avoid is the implication that society has become utterly involved in the workings of research, and that every scientist who wants to perform any kind of experiment must first explain its environmental impact. Whether we like it or not, we are going to have to trust some level of scientific intuition; we cannot screen every type of proposed scientific advance. Somewhere along the line, differentiations must be made so that the scientist involved in constructing a revolutionary rocking chair is not subject to the same pressures as the one playing Sorcerer's Apprentice with nature.

While it is true that examining the problem of society's powers of regulation over science from this perspective leaves practically all of the initial stages of regulation in the hands of the scientists themselves, it seems impossible to imagine how else the problem can be resolved within the confines of the soceity in which we now live. Big Brother tactics notwithstanding, there is no practical way that the public can force the scientific sector to reveal the true nature of its work—and thereby to weaken its own control over its processes—unless it demonstrates flexibility and consideration. On the other hand, it can and must demand that the scientific community set up regulating bodies on a more formal basis than now exists to give researchers official conduits through which they can channel both inquiries and complaints. Science, with the capability to change the face of the earth through any one of several disciplines, is a natural power structure; like any power structure, it must develop a system of checks and balances analogous to those already overseeing the economic, political, and military branches of our society.

It is not quite enough, however, for us to simply sit back and wait for scientists to ask us to help resolve prob-

lems which they consider to be within our purview. Before
any dialogue begins, we must join with them in preparing
a set of guidelines which defines exactly what does and
does not constitute ethical research.

In a vacuum, scientific research can be defined simply
and clearly as a quest for information. Within such a definition, ethical conduct consists of any work that strives to
expand the frontiers of knowledge. Unethical conduct includes research that strives to turn science away
from truth—research, for example, that manufactures
knowledge to support an otherwise untenable hypothesis
(as was the case with Lysenkoism in the Soviet Union).
The two sides to the internal ethics of scientific research
are, indeed, quite simple.

But the ethics of research grow considerably more complex when external factors are brought into consideration.
Those factors of moral policy and social ethics expounded
by Dr. Callahan are also crucial concerns. In their article
"Toward a Theory of Medical Fallibility," Samuel Gorovitz and Alasdair MacIntyre provide a striking example:

Those doctors who performed experiments on living prisoners
in Auschwitz did not violate any of the internal norms concern-
ing truth-seeking and problem-solving. Indeed, at least some of
them might have been quite exceptionally devoted to those
norms. . . . The scientist discovers what he or she can. It is his
or her duty to pursue empirical truth, but as a scientist he or
she has no further concern with the social effect of the discoveries or the ethical status of the process of inquiry that led
to them. As citizen, as parent, as teacher, he or she must have
moral concerns. But this will always be a matter of norms external to science.

Nevertheless, scientists do not exist in a void while they
are performing their research. Whatever their adherence is
to the internal norms of science, the question of their
place in society persists:

Hundreds of years of understanding nature as a mere presence
where particulars are nothing more than specimens for study
. . . have contributed strongly to the ecological violence which
we have done to nature. That is, not merely the forms of our
economy or our technology, but also—perhaps surprisingly—

those of our science have contributed to our estrangement from nature and from other species. To say this is in no way to decry experiment, the search for laws, or the construction of fundamental theory. It is to say that the norms which are internal to the project of understanding nature . . . turn out to be broader and more complex than has generally been acknowledged. The S.S. doctors were indeed violating a relationship to men and to nature which is an essential part of understanding men and nature; they thus failed as scientists and not only as men and citizens.[4]

This, then, is the area of potential conflict. The scientist is faced with a fundamental dilemma, a possible clash between the demands of scientific ethics and those of the public, between what research finds ethical and what the public deems acceptable. It is the obligation of the public to avoid embroilments on the level of the XYY controversy as best it can and to leave to the scientist a major part of the responsibility of deciding the technical aspects of what is and is not risky for the public.

Society must also devise a system to replace the *ad hoc* decision-making process that has characterized the recombinant DNA debate, the XYY controversy, and other similar issues. As we advance beyond the Industrial Age, we must realize that history gives us predictive powers. We no longer have the luxury of simply reacting to every crisis that falls upon us.

What Science Owes Society

The psyche of a scientist is an extraordinary thing. The rigors of his discipline are so focused that it is as difficult for outsiders to understand many of the scientist's motivations as it is to comprehend the scope of his work. With its emphasis upon concentrated, solitary work, with its demands for answers that reduce the greater questions of the universe to their fundamental physical states, science requires its practitioners to live differently from the ordinary citizen. Whole professional careers are spent in intense ef-

---

[4] Samuel Gorovitz and Alasdair MacIntyre, "Toward a Medical Fallibility," in H.T. Engelhardt and D. Callahan, eds., *Science, Ethics, and Medicine* (Hastings-on-Hudson, N.Y.: Institute of Society, Ethics and the Life Sciences, 1976), pp. 260-62.

forts to track down ever tinier secrets. And some place along the line, many of the original goals of research have been forgotten. The questions that science has been trying to answer since science began are now buried in mountains of information. Research has become a truth in its own right. And facts are the pots of gold at the end of the rainbow. As Albert Schweitzer once observed, "Our age has discovered how to divorce knowledge from thought, with the result that we have . . . a science which is free, but hardly any science left which reflects."

The scientist's preoccupation with his work is legendary. One young molecular biologist at the Asilomar Conference, when asked by a reporter to describe the sea urchins whose genetic structure he had been working on for years, responded, "The only thing I know about sea urchins is what I see under a microscope." The animal no longer meant anything to the scientist; its chromosomes had become everything.

It is obvious that perceptions such as this must breed distorted views of the outside world and society in general. And there is little question that this one scientist is not alone; many in science consciously avoid the social implications of their research. Indeed, many have entered science at least partly to escape the realities of the everyday world. And while they may consider talk about social and political consequences, ethical standards, or problems of life and death interesting, worrying about such things is like being on vacation. The work is all-consuming; with the level of sophistication needed for advancement, and with the competition that modern methods of research have spawned, it has to be. But it means that many scientists live and work in a world divorced from the realities of society. It means that the internal ethical norms of research are often the only ones that matter, and that many scientists view themselves as an elite, separate from society and somehow above answering to it. Unfortunately for these scientists, while their decision to turn their backs on society's ethical problems may be their prerogative as individuals, when their perceptions interfere with the working relationship between science and society, the public must become concerned. For the bottom line of that relationship is that society can, ultimately, survive without science, but science cannot exist without society.

The subordination of science to society is as important to our system of social realities as the cell is to biology. Scientific freedom of inquiry is simply not as fundamental a right as the public's right to its safety. The covenant between science and society implies that research must be unfettered as long as it is safe. But if the public senses danger in a particular line of research, it has the right to debate the value of the research; and if that line proves too risky in the public's judgment, the public has the right to stop it. In the end, science is as responsible for the public welfare as the government, the fire department, and the licensed driver.

Because scientific research is an active, driving force, the responsibilities of science to society demand action. If the public must keep its hands away from research until a red flag is waved, the scientific community must be ready and willing to wave that flag whenever the potential for hazard is substantial. This means that each scientist is not only responsible for the safety of his own basic research, but for investigating the foreseeable consequences of that research. It means that the researcher's actual responsibilities extend beyond the scope of his own work to the results that others might achieve through the use of his findings.

In recombinant DNA research at least, instances of foreseeable problems are not hard to find. In the light of how we used our past breakthroughs in physics, it is easy to see how the potential of biological warfare has been enhanced by the new techniques. Even though the United States, the Soviet Union, the United Kingdom, and over a hundred other countries have signed the Convention on the Prohibition of the Development, Production and Stockpiling of Bacteriological (Biological) and Toxin Weapons—a treaty in effect since March 26, 1975—it would be unwise for us to underestimate the monumental power of international suspicion and the possibility of future attempts to circumvent or even ignore the treaty's provisions. Already two articles published in the Soviet Union "indirectly but clearly argue that genetic weapons are new and outside the scope of the bacteriological treaty."[5] Indeed, since the treaty itself states that it prohibits only those microbial agents "that have no justification for prophylactic, protective or other peaceful purposes," even the least imaginative government official would have little trouble finding a way to make research into defensive biological weapons (and related research) fit into its provisions. In the light of the United States' own history of collaboration between the military and the universities (the Army has reported that eighty-eight institutions conducted research into biological warfare under more than 300 defense contracts in the years 1942–71), there is little doubt that funds for such research would find eager takers in molecular biology.

Controversial applications other than biological warfare may include the controlled alteration of the human genetic structure, sophisticated asexual reproduction (the less than attractive possibility of what has been called "wombs for rent"), and various radical forms of genetic surgery. While many of these advances may be beneficial to society from a technological standpoint, the ethical questions they raise cannot be ignored. And researchers must understand that they are responsible for presenting the facts even though the final decision is not theirs to make. Scientific expertise does not qualify scientists as arbiters of the merits or hazards of the societal impact of their work. In the end, they are nothing more than equal members of society. The ultimate power to make these decisions lies in the hands of those who will have to carry the burden—the public—and not in the hands of the technicians.

If science had ever actually attained the level of impartiality that many of its members have claimed, things might be different. But history has exposed countless episodes of unethical scientific practice performed under the guise of research. Quite recently, newspaper stories chronicled the agony of a group of black prisoners in Alabama who, in the 1930s, were left with untreated cases of syphilis so that their doctors could chart the course of the disease. Other social and political uses of science in the past have confirmed the conviction of many that while science in its purest form is a search for the perfect truth, it is still

---

[5] *Los Angeles Times*, February 22, 1977.

performed by imperfect human beings, scientists, who are subject to the same prejudices, the same distorted priorities, the same illusions as everyone else.

The fallibility of scientists surfaces daily. But it becomes most obvious in some examples of ideological science—like the eugenics movement—that have periodically afflicted us throughout history.

The pursuit of eugenics, the science of genetically improving the quality of the human race, actually began in the nineteenth century. While it surfaced as a legitimate branch of the rapidly developing discipline of biology, eugenics soon became a catchword, a useful way of explaining some of the more bizarre social policies that needed scientific validation. In its strictest sense, the original concept of eugenics is admirable; after all, who would not agree that we would be better off if such genetic deficiencies as hemophilia and Down's syndrome were removed from the human gene pool? But, in the world of the late nineteenth and early twentieth centuries, "genetic deficiencies" were interpreted mainly to include characteristics inherent to whole classes of people, classes that shared the single, unifying feature of being repugnant to those in power.

The rediscovery of Gregor Mendel's observations in 1901 brought with it what seemed to be strong scientific support for the entire concept of eugenics. Single genetic differences had been shown to control major physical characteristics in humans as well as in plants and lesser animals; applying the same laws to human behavioral traits seemed a logical extension of recognized fact. So it was that many of the leading geneticists of the day agreed that defective genes were the determining factors in a number of socially unpleasant conditions. They concluded, for example, that "alcoholism, seafaringness, degeneracy, and feeblemindedness were each due to singe mendelian factors," and that mulattos would have "the long legs of the Negro and the short arms of the white, which would put (Why they did not consider the opposite—which would seem to have given mulattos a decided genetic advantage—was never explained.) Eugenics thus gave a scientific veneer to the argument of the racists of the early twentieth century.

them at a disadvantage in picking things off the ground."⁶

In 1924, the eugenics movement in the United States capped its political efforts by lobbying successfully for the Immigration Restriction Act. By basing quotas of immigration into the United States upon the ratios that existed in 1890, the Immigration Act set limits on the massive influx of southern and eastern Europeans, which had begun during the first two decades of the twentieth century; in 1890, the majority of immigrants came from western Europe, from populations far closer to the genetic construct that the American power structure considered ideal. Both the testimony preceding the passage of the Immigration Act and the act itself were blatantly racist in content; yet the law stayed on the books until 1962, when Congress finally voted in a new, more equitable law.

Although the strength of the eugenics movement began to wane soon after the passage of the Immigration Restriction Act (corresponding to the rise in power of another, far more explicit policy of eugenics in Nazi Germany in the 1930s), its influence continues to distort state and local social policies even today:

Between 1911 and 1930, 33 states passed laws requiring sterilization for a variety of behavioral traits deemed to be genetically determined. These included, depending on the state, such characteristics as criminality, alcoholism, tendency to commit rape, sodomy, or bestiality and feeblemindedness. Many of these laws are still on the books and have resulted, since that time, in at least 60,000 sterilizations. For instance, it has been reported that the Eugenics Board of the state of North Carolina sterilized 1620 persons between 1960 and 1968, mostly young black women. By far the greatest category of sterilization under these laws was for feeblemindedness. Feeblemindedness can supposedly be defined by the administering of an IQ test. To get some idea of the validity of IQ tests in this regard, we need only look back at the results of Goddard, who was asked in 1912 by the United States Public Health Service to use the tests to determine the frequency of feeblemindedness among new classes of immigrants into the country. Goddard's results demonstrated that

⁶ Quoted in Jonathan Beckwith, "Social and Political Uses of Genetics in the United States: Past and Present," in M. Lappé and R. S. Morison, eds., *Ethical and Scientific Uses Posed by Human Uses of Molecular Genetics* (New York: New York Academy of Sciences, 1976), p. 47.

the following frequencies of feeblemindedness pertained: among Hungarians, 83%, Russians, 87%, Jews, 83% and Italians, 79%.[7]

It is not only the danger of this kind of prejudice and the possibilities of distorted scientific vision that make it necessary for scientists to bow to some form of societal control. Even the purest of scientists have been suspected of fudging experimental data for the sake of their theories. A case which shakes the very foundations of the scientific Olympus is that of Gregor Mendel and his pea plants. While Mendel's paper claimed that he had observed his ratios of dominant-to-recessive genes in succeeding generations of his plants, the odds that he actually did so have been calculated at about 10,000 to 1. In truth, Mendel's ratios are nothing more than mathematical probabilities. They have the same predictive values as flipping coins or rolling dice. In a coin flip, the fifty-fifty chance that a head or a tail will appear in any single flip does not mean that two flips will exhaust the possibilities. Rather, it means that factors of chance will balance out in the long run and that—over a period of 1,000 or 10,000 tests—approximately 50 percent of the flips will come up heads and 50 percent tails. Mendel may have observed a certain pattern in the "coin flips" of the recessive and dominant genes in his plants; he may have calculated what their ratios would be over a period of time; but he almost certainly did not actually observe three-to-one ratios of dominant to recessive genes in each second generation. Much of his "empirical evidence" was probably buried deep within the imaginative recesses of his mind.

Scientific obligations to society thus actually begin with the fallibility of scientists themselves. The problems of controlling dangerous or unwanted knowledge only reinforce their basic responsibility. Scientists may represent mankind's attempts to achieve the ultimate levels of objectivity and impartiality that are a part of scientific truth, but they are no more the embodiment of pure science than politicians are of democratic government.

Nevertheless, we have always been reluctant to police

7 Beckwith, p. 48.

the activities of our technicians, both before and after scientific breakthroughs have brought us significant new hardware. As a society, we have learned to react to the presence of danger only after it is present. While this might be sufficient in, for example, the problems of nuclear technology (although it is doubtful that problems like storing nuclear waste and the safety of communities situated near breeder reactors will be resolved before an ocean is swept clean of fish or a city is leveled), the acceleration in scientific technology is bound to catch up with us in the end. Sooner or later, we are bound to be confronted with knowledge that we do not want. But unlike for Galileo, the threat to us will not be philosophical or theological, but physical.

The final scientific obligation will be the hardest for researchers to swallow, for it implies that regulation will be based on speculation rather than on demonstrable risk. It seems far more reasonable for society to claim that ongoing research is actually violating its rights than for it to inject imagination into a debate and demand that speculation carry weight. But as science grows more powerful, the checks upon it must also grow. And if society recognizes its inability to slow the pace of dangerous research once it comes into existence, science, as a part of society, must share the responsibility of predicting and curtailing such research.

It is obvious that no legislation will be subtle enough to balance the legitimate rights of science against the speculative fears of society. Such a balance, and the constraints that result, must be etched in the very spirit of the relationship between the two. While an embryonic understanding has developed during the debate surrounding recombinant DNA, we must ensure that it matures in the future.

Public Participation in the Recombinant DNA Debate

The potential risks and benefits of recombinant DNA technology should make it clear that science and society are not merely entering an era of struggle, a contest of parry and thrust where superior power wields the upper hand. We are moving toward a time when agreement is crucial to our very survival, when the real struggle is between our will to survive and our power to destroy.

The basic guidelines that science and society must follow to establish a constructive relationship are partly a matter of our following our instincts, partly a matter of our reflecting on and setting an ordered course for the future. One of the keys to success will revolve around the degree of public participation, around the extent to which the public voice is heard and heeded in the decision-making process.

In the debate surrounding recombination, there is little doubt that the public voice was heard. But it is questionable whether it was heeded. For it is not the outcome of the debate but its process which must guide us in setting up our future course; it is not the end but the means which is a predictive factor in future controversies.

The controversy began the way most scientific controversies will begin in the future: with scientists taking a hard look at a line of research which seemed to contain danger as well as promise. The first stirrings of unease were purely scientific; Robert Pollack made his telephone call to Paul Berg, Berg discussed the issue with his colleagues, the Gordon Conference expressed its concern, and the MIT Eight composed their letter to *Science*, the National Institutes of Health, and the National Academy of Science. Then the storm broke. With the publication of the Berg letter, the public gained its first substantive knowledge of recombination and its hazards.

Because the scientists involved still saw the issue as one of technology—of controlling the accidental release of novel organisms—they continued to limit the extent of the debate. The Asilomar Conference was convened as a closed gathering, limited to the scientific cognoscenti; the committee constructing the NIH guidelines held its private meetings without the distractions of the public's priorities. Only in February 1976—when the guidelines had been given tentative approval, when the debate had existed for almost three years—was the public invited to voice its concerns to the NIH advisory committee. While that meeting brought out perspectives that had been ignored during the long months of preliminary decision-making, few of the ideas presented in opposition to the proposed guidelines were incorporated into the final documents. The guidelines—the culmination of the initial decision-making process surrounding recombinant DNA—were almost exclusively the product of scientific concerns, scientific priorities, scientific definitions of benefit and risk.

It was only after these hearings and conferences that the public voice in the recombinant DNA controversy rose in volume. Ironically the day which marked the first flutterings of public influence—June 23, 1976—was the same day that the NIH guidelines were published. On that day, the City Council of Cambridge, Massachusetts, met to discuss the impending construction of a P3 laboratory at Harvard University, a laboratory that was to house recombinant DNA experiments.

The hearings of the Cambridge Experimentation Review Board (CERB) were the first that pitted advocates and critics of the research against each other in front of a jury of the public. During the six months of local debate, scientists had to undergo an entirely new experience: that of sitting in front of a panel of laymen and being asked to account for the social and ethical implications of their work. While the major portion of the board's work concerned the immediate biohazards of the research, and while the board's final conclusions were based primarily upon the methods proposed by Harvard to ensure the safety of the Cambridge citizenry, the very acceptance by the researchers of the CERB's role as reviewer of their proposed work marked a startling and radical shift from past proclamations of science's immunity from external interference.

The board took its role seriously. Composed of nine members—from a structural engineer and two nurses to three people long active in Cambridge politics—the group had no close identification with either MIT or Harvard. Its stated goals were tied to the interests of the Cambridge community as a whole. In determining the boundaries of the problem, the board was faced with a crucial dilemma: none of its members knew anything about the research that it was to judge. But it went about solving this problem in a businesslike, logical way:

- Each Board member was provided with special technical documents on the controversy, including the NIH guidelines . . . [and essays from the technical and popular press].

- A technical assistant to the Board, who had training in the biological sciences, offered help with translating technical concepts. . . .

- Spokespeople who appeared before the Board were asked to reduce technical concepts to layman's terms, to present simplified models of biochemical events, and to draw upon analogies whenever they were available.

- The members of CERB were witness to a forum on the . . . controversy. . . .

- Two open telephone conversations were used to draw testimony from people outside the state. In one of these conversations, the Director of NIH and a panel of experts responded to questions of the Board members.

- In a five-hour marathon session, CERB carried out a type of mock courtroom affair. Board members served as a . . . jury, while advocates on both sides of the issue presented their case, were given an opportunity to cross-examine one another, and responded to questions raised by the "citizen jury."

- Board members were taken through laboratories at Harvard and MIT. In one case a mock experiment was carried out which exemplified the various stages of the recombinant DNA process.[8]

In the end, CERB's recommendations closely approximated those of the NIH. But it was the process, as well as the result, that mattered. Like the regents at the University of Michigan before it—and committees in almost a dozen other locales after—CERB showed that a conscientious ef-

[8] Cambridge Experimentation Review Board, "Guidelines for the Use of Recombinant DNA Molecule Technology, in the City of Cambridge," January 5, 1977, pp. 10-11.

fort on the part of citizens empowered to oversee the debatable aspects of scientific progress would be enough to give an issue a fair, complete hearing. Most critics and advocates agreed that the board's decisions were based upon legitimate criteria, that the process initiated by the Cambridge City Council had been impartial and open, and that the results reflected the actual concerns and interests of the community. Not everyone agreed with the board's decisions—in fact, Mayor Alfred Vellucci, who was the driving force behind the public debate and decision-making process, claimed that CERB had been pressured into constructing relatively lenient guidelines by the suasion of the scientific community. But its decisions now seem as reasonable and logical as any that might have been developed by the scientific community alone.

The shift to local decision-making marked the final stage of debate in the controversy surrounding recombinant DNA. From a purely scientific discussion, it had finally evolved to the point where power no longer lay in the hands of the technicians. What the organizing committee for the Asilomar Conference had originally viewed as a purely technical issue that lent itself best to scientific debate had now come full circle. Even though the issues were still technical, the layman was involved. Through a long process of struggle, the public was finally represented in the process of decision-making. And its power evolved in a strikingly traditional way, not from the exercise of the centralized influence of the federal government, but from smaller centers of concern, from citizens who felt that they were being directly threatened by the new technology, from what amounted to a series of local grass-roots movements. Congress was still struggling with forms of legislation that would placate lobbying efforts on both sides of the issue; states were fighting similar battles; but local communities like Cambridge, Ann Arbor, Michigan, and Princeton, New Jersey, had recognized that the potential threat was directed at them, and had gone ahead and done something about it.

That most communities decided in the end that the NIH guidelines were, in the main, adequate is beside the point. The key to what occurred lies in true public participation; local communities literally wrested control of the issue from the national agencies who had been making what amounted to unilateral decisions and made the responsibility for safety their own. Once they had done that, and once the scientific community had acquiesced, they had come a long way in changing the future relationship between themselves and scientific research. Arguments as to who the public was and whether Congress or other legislative bodies had the power to control the progress of science had become superfluous. By acting while others argued, the public had shown its real strength.

The issue of recombination is, of course, far from over. It will take decades before all the problems and risks have been sorted out, before we can find out whether the research *is* really safe, before questions about scientific accident and the potential of deliberate misuse of our new powers are answered. But the course of the debate over recombination during the past few years is clear enough to reveal some broad truths about the most effective methods of communication between science and society:

- The way that the DNA debate was initiated—through a spontaneous, *ad hoc* response by concerned scientists—is the most natural, efficient, and logical way for a potential problem in science or bioethics to gain attention.

- The ripple effect of the debate—growing from a small, technical problem better discussed by informal gatherings of scientists to a major topic of concern for a large, public meeting of experts—is also a natural one. As the dimensions of the problem became clear, its audience grew. The debate indicated that it would be disastrous to treat every potential controversy as if it were a live bombshell from the beginning; constant doomsaying would not only distort the differences in risk between problems, but would desensitize the public to real dangers and compel scientists to keep quiet about their more controversial projects.

- The decision-making process itself, which in the current debate remained the province of those in

power, will require stricter regulation and control
in the future. Mechanisms that include all responsible parties in the process and that make the final
decisions applicable to all researchers and facilities
must be established. As it now stands, industrial
and pharmaceutical concerns working with recombination still have no legal obligation to adhere to
even the slender restrictions of the NIH guidelines.
And although most of those companies claim to
have taken the guidelines to heart, the public is expected to rely upon little more than their spotty
record of past honesty as proof of their intentions.

● In the future, it will not be enough for any ideological or technical sector of society to claim that a
series of well-publicized debates in diverse locations, media attention, and the protests of public
interest groups constitute adequate representation
of the public welfare in a serious debate.

In a way, we have been lucky that recombinant DNA
research was the first in what promises to be a long line of
controversies testing the bonds between science and society. Recombination has lent itself to a slow exploration of
the process of making decisions. The results have probably
not placed us in too terrible jeopardy, since the evaluation
of risks and benefits has led scientific and public forums to
similar conclusions, while the more complex questions of
application of research lie far enough in the future to give
us time to establish the necessary mechanisms of decision-
making and control. It is absurd, however, to argue that
the debate over recombination has involved the public in
the way that an issue of such importance should. The public pushed its way into the process late in the day, at a
point which might have been meaningless had the issue
had more immediate repercussions.

The various stages of the recombinant DNA debate
have made it clear that it is time for us to recognize how
precarious our position is in the world today. Research is
not only a constant process but an ever-expanding one.
Science is moving forward on all fronts, and while it is bi-

ology that is now heralding a revolution, other sciences can easily do the same. Our acquisition of power has accelerated to the point where we must either learn to control it, or have it control us. As Albert Rosenfeld of the *Saturday Review* has noted, "We have not respected any region of knowledge as no-man's land. We have not forbidden ourselves fruit from any tree. The ironic result is that we may have to take upon ourselves some godlike prerogatives as we become self-anointed trustees of our own evolution."

SIX:

Resolving the Controversy: What Can Be Done?

In resolving the problems that recombinant DNA research has posed, we are really faced with two separate dilemmas. First, now that the risks of recombination have been exposed, how are we going to deal with them? And second, how are we going to create practical mechanisms to resolve similar controversies in the future?

The practical problems of protecting ourselves against the risks of both basic and applied research into recombination are monstrous, given the nature of the research, science, and the international community. Alone, each would make practical regulation a formidable challenge. Together, they present us with a set of obstacles that are almost insurmountable.

Recombination is, of course, a tremendously accessible line of research. The enzymes, vectors, and hosts needed to perform the experiments are readily available, with no questions asked, from several biological warehouses. And while it is highly unlikely that our brilliant high school biochemist will indulge in the fantasy of recombination, it is quite possible that someone else with the necessary knowledge will play with the techniques. If that happens, there is literally nothing that either the scientific community or the public will be able to do. As Erwin Chargaff recently noted, "Science is poorly equipped to police the sick imaginations of some of its practitioners."

Just as we must recognize that we are powerless to do anything about aberrant research that may be undertaken in private, we must also acknowledge our inability to supervise international recombinant DNA research. Although political boundaries effectively prevent us from exporting even the most reasonable guidelines without the active consent and support of other nations, the same boundaries cannot stop the spread of anything infectious that might escape during a laboratory accident in some distant locale. While discussing the expansion and sophistication of technology in the 1960s, Marshall McLuhan noted that, at least for communications, the world had become a global village. With the advent of the jet age, the same is now applicable to disease, for an escaping microorganism in West Germany, Switzerland, or the Soviet Union is little more than a day's travel from the United States. In a way, advocates of restricting the research are caught in a practical bind. The only way to provide effective regulation in the long run is to back up reasonable guidelines with the teeth of significant penalties, while the only way to gain international compliance is through voluntary agreement on the part of the governments involved.

All these factors seem problematical at the moment. It is still easy for governments (and industrial concerns) to agree to regulate research tightly. But all too soon, another factor besides altruism and the excitement of discovery will descend upon the present debate. And that factor—money and the possibility of profit—will sorely test the promises that both industry and other countries have made to adhere voluntarily to the present guidelines.

Industry and Recombination: Will We Be Told the Truth?

The techniques themselves are relatively inexpensive; the theoretical possibilities for profit—from mass production of important biological products like growth hormones to the growth of massive vats of metal-hungry microbes that might be used to mine valuable trace metals from the oceans—are immense. There is indeed gold in them thar bugs. And it will take only the transition from vague future possibility to immediate probability to bring the money vultures out in force. If past experience means anything, their appearance will toll a death knell to questions of ethics and safety in recombinant DNA research.

Already industry is involved to varying degrees. Most of the work is being done in the pharmaceutical field by such giants as Miles Laboratories, Eli-Lilly & Company, Hoffman-LaRoche, and the Upjohn Company, but other industrial stalwarts, like General Electric, Monsanto, and Du Pont, are also performing or supporting research. If these and other established concerns were the only companies exploring the potential of recombination, we would probably have little to worry about, for their position in industry and their tremendous sensitivity to the pitfalls of public emnity practically ensure that their emphasis upon safety would equal or exceed that of university laboratories. But the esoteric recombinant experiment of today is the oil fever of tomorrow. Constructing an adequate laboratory and hiring a handful of hungry young scientists is within the limited capabilities of even the most hard-pressed industrial gambler. And fly-by-night corporations with names like Clean Genes, Inc. and Buy-A-Babe are almost as certain to appear as tomorrow's weather. Already, since the successful creation of bacteria that can manufacture insulin, several scientists in the San Francisco area have formed their own company and have applied for patents on both the process of recombination and the money-making, insulin-producing microbes themselves. And while the scientists involved are respected researchers from Stanford University and the University of California and are expected to perform their experiments as they have their careful scholarly work, others might not. Worse, until Congress passes a law that extends the provisions of the NIH guidelines to industry and makes dangerous experimentation in recombinants a federal offense, they will not have to. For there is not a single federal statute or law which covers the conduct of recombinant DNA research.

The likelihood of meaningful federal legislation on industrial use of recombinant DNA procedures is, at best, slender. The problem is one of practical political power; the industries that would feel the effects of federal legislation most severely are the same ones whose lobbying contingents on Capitol Hill wield vast and pervasive influence. Any legislation that is permitted to pass without extensive opposition will have loopholes large enough to accommodate a P4 laboratory.

Several bills have been introduced in both the Senate and the House of Representatives, but the pressures brought upon the legislators have been so great that the various pieces of legislation have been either shelved or sent back to committee. The bill considered to have the best chance of passing, proposed by the Carter Administration and introduced by Senator Edward Kennedy, would have given the power to license and register recombinant DNA facilities and projects to the Secretary of Health, Education, and Welfare. Violators of its provisions would have been subject to stiff civil and criminal penalties. Local statutes would have been preempted—unless they could be shown to be tougher than the provisions of the federal legislation—and the entire country would have been subject to a uniform set of regulations, a single standard of conduct.

Senator Kennedy's political instincts, however, are notoriously sensitive. By October 1977, his recombinant DNA bill had been withdrawn from the calendar; it was announced that new information, which indicated that the dangers were not as great as had once been feared, impelled the Senator to reexamine his proposals. In reality, the combined might of the scientific and industrial lobbies have probably killed the bill; and although it is expected that some kind of legislation will be passed within the next few years, there is little chance that it will amount to much more than the NIH guidelines, modified slightly to suit the special needs of industry.

There are, of course, serious questions as to the worth of legal controls even if they are passed by Congress. The creation of a new bureaucracy to handle not only the supervision of recombination but the vagaries of other research of questionable safety is bound to cause significant disruptions for safe science, as well as for those projects that require regulation. Bureaucracies are not known for their flexibility; neither have they always received the highest marks for effectiveness. They are, in short, unreliable.

The welter of bureaucracies eligible to manage the regulation of recombinant DNA research introduces the additional problem of jurisdiction. Right now, candidates for the job are scattered all over the landscape. In addition to a new bureaucracy, they range from the Center for Disease Control in Atlanta, Georgia (which regulates interstate shipment of hazardous biological agents) and the Food and Drug Administration, through the Patent Office (which must evaluate applications for control of new processes and new biological forms alike) and the Environmental Protection Agency, to the NIH, which, in the recent past, has already demonstrated its reluctance to regulate even the research for which it provides funds. Of all the candidates, the least attractive from the standpoint of regulation must be the NIH. Its own past record, and the records of other agencies that have been asked to control the very research they are supporting financially, demonstrates that the NIH would have serious difficulty in providing reasonable supervision.

The history of the Atomic Energy Commission is a case in point. Created to sponsor and support atomic energy research, the AEC also became responsible for its regulation. Its position meant that it had control over funds for research and development, as well as a voice in which nuclear reactors would and would not be built. Its conflict of interest did not sit easily with the commission, and it wore its two hats with great difficulty. The problem worsened in the late 1950s and early 1960s. Because nuclear energy was considered to be the technology of the future, hundreds of nuclear power plants were planned for the United States. The AEC found it impossible to encourage the expansion of nuclear power while protecting the public; ultimately it practically abandoned all semblance of responsible control. The results were almost catastrophic. The uninvolved sectors of the scientific community (in particular, the molecular biologists) were up in arms against what they considered a dangerous lack of concern over a type of technology which contained proven dangers. Several times, their warnings and predictions of doom came uncomfortably close to reality. In one instance, a breeder reactor on the shores of Lake Michigan was shut down just seconds before it would have exploded and annihilated the city of Detroit. Other accidents and near accidents were reported from the Dakotas to Alabama, and an actual explosion was rumored to have occurred in a power plant in the Soviet Union. Only one of the American incidents actually took lives, and the AEC and other governmental agencies effectively cut off the publicity that these borderline disasters deserved. But the fact remains that, while the AEC was not responsible for many of the problems that the nuclear energy industry was having, others were directly under its control. In several instances, the lack of supervision in the construction or operation of a reactor was isolated as the primary cause of the trouble.

The NIH is in a position in recombinant DNA research similar to the AEC's in nuclear power, for, as a principal source of funding and regulation, the NIH must somehow both encourage and restrict research. The pressures upon the NIH to regulate as little as possible are huge. First, its very existence is predicated upon breakthroughs in biomedical research, and not in preventing accidents. Its staff works primarily to distribute money. Its success or failure is measured not by the number of accidents it prevents, but by the number of strikingly successful projects it has underwritten. And if those projects provide breakthroughs that are meaningful in the fields of both medicine and public relations, so much the better. Like many governmental agencies, the NIH depends at least partially upon its public reputation for increasing its funding. A Congressman who can return to his home district during an election campaign and claim to have sponsored a bill that resulted in the development of an artificial heart or something equally dramatic will be far more kindly disposed to renewing or increasing biomedical funding than he would if his support for research has resulted in nothing more significant than a dearth of traumatic accidents. Whatever the criteria for success might be within the halls of the NIH, success for a government agency may be freely translated as actions which generate Congressional approval.

Confusing the issue still more is the fact that the industrial lobbyists do have valid points in their battle against restrictive legislation. Legislation that would merely make the NIH guidelines applicable to all forms of recombinant DNA research, they note, would be completely unrealistic, since many of the guidelines, designed specifically for academic environments, do not take into account the special problems of industrially oriented research. The limit, for example, on ten liters of culture for any recombinant work would obviously not work if industries were to use recombination as a tool for profit. And the requirement that all projects be outlined and publicly revealed before they are approved would prevent companies from patenting their technology and creations, since the patent law requires that research plans not be disclosed. Furthermore, the companies that are large enough and strong enough to have lobbyists in Washington are those least likely to violate intentionally the applicable guidelines of the NIH; they have records of safety in research which compare favorably over the years with those of the academic research community. Legislation will actually be aimed not at these giants but at the quick-money research factories that are bound to spring up overnight; yet any legislation aimed at them would also affect those who have shown a degree of restraint in the past. On the other hand, if legislation were weakened in deference to the past records of the industrial and pharmaceutical giants, the laws would serve as engraved invitations for anyone who wanted to buy the necessary chemicals and play God in his basement to act without fear of legal retribution.

In legislation, as in the guidelines first proposed by the NIH, the problem all along has been to determine who should decide. With the national character of the research, and with risks rising from both basic and applied research affecting people on a national, if not global, scale, regulation theoretically falls in the lap of Congress. But in practical terms it is almost more important to answer a second question, whimsically asked by the social scientist Donald Michael: Who decides who decides? Thus far, the answer lies with the directors of the research, with the companies that can bring heavy pressure to bear upon the legislature. For it is those who will profit by the research—whose perceptions must be somewhat distorted—who wield the real power. The only control the public has is the companies' recognition that adequate regulation might help them in the long run, since an accident—and the public outrage which would inevitably follow—would be far more harmful than even the most burdensome regulation. And that kind of control is tenuous at best.

The question therefore remains: Because the danger of an accident is so slim, and because regulation is so costly that it would seem economically clever for companies to take chances, will we be told the truth? Without adequate legislation or the means to regulate techniques that can easily be abused, odds are that we will not. The present system of regulation is perpetrating a dangerous double standard, under which researchers in the academic community are held responsible for the foreseeable results of their work, while scientists in industry may experiment with relative impunity. The reasons for this dichotomy include our basic respect for the free marketplace and our *laissez faire* attitude toward privately funded research. But the results of experimentation are the same. Organisms escaping from the experimental laboratories of Hoffmann-LaRoche are no less hazardous than those skipping ship at Harvard. With all the concentration focused upon the obligations of science and society, it is ironic that a double standard based not upon the preeminence of the scientific elite but on the value of money might ultimately do us in.

As a guiding example of regulation, the issue of recombination thus sets a poor standard. Every time a problem seems solved, another crops up in its place; every time our attention is focused upon one aspect of the situation, others sprout in profusion. The entire issue, if diagrammed, would look something like a spastic spider, with its legs out of control and walking in different directions.

If the issue of recombination has shown us anything, it is that we cannot afford to go through future crises with such a lack of preparation. In addition to reaching fundamental agreements on moral policies, science and society must construct mechanisms for carrying them through, forums in which arguments of fact and fiction can be separated and then passed on to the legislative and regulatory bodies for action. Several such forums have been suggested. One of the most promising candidates is a quasi-legal hearing, a court proceeding modified for science.

The Science Court

The Science Court was first suggested in the late 1960s by Dr. Arthur Kantrowitz, a physicist who had helped solve the problem of ballistic missile and spacecraft reentry years before. Since then, his concept has been supported by such organizations as the American Chemical Society, the American Physical Society, the American Society for Microbiology, and the American Association for the Advancement of Science. In a joint declaration, these and other such organizations declared that the Science Court "offers a potential mechanism through which the status of knowledge or lack of knowledge on a controversial issue could be clarified in open forum as an input to the policy-making process."

The Science Court would act as a bridge between the scientific community and uninformed policy-makers in government by providing an institutional setting in which adversary proceedings could be held. With perhaps five to seven judges drawn from the ranks of scientists and laymen alike, the court could offer a respected, influential mechanism to anyone wishing to question the controversial aspects of any line of research and development.

The strengths of such a forum are obvious. With the power to subpoena witnesses from both sides of an issue, with an impartial board sitting in judgment over an inquiry analogous to a legal proceeding, the Science Court could become an important fact-finding body, acting in an advisory capacity to legislators and government administrators alike. In addition, it would provide a logical place to turn for advice about scientific issues that have outstripped the limits of their technology and have begun to infringe upon accepted scientific and social standards; indeed, it would have been a perfect place for the debate over recombination to have been held after the conferees at Asilomar had shown themselves incapable of dealing with the greater issues implied by the research. Because its role would not include issuing verdicts but would be limited to producing advisory reports upon the current state of technical knowledge, it could act as the clearing house for factual disputes that seem to color most scientific debates; it could record the points upon which both sides agree and render judgments on those that are in dispute. Ultimately, as its reports are published, it would act as an informational tool for the public and legislature alike, superseding the myriad methods of trial and error that seem to be the state of the art in scientific debate and providing a unified, unbiased viewpoint. The sole responsibilities that adversaries in such a setting would have, would be to agree upon the judges, choose their spokesmen, and organize their cases. The other factors so tightly bound to decision-making today—the media, informational lobbying efforts, and the like—would prove unnecessary.

The Science Court is, of course, no panacea. But its weaknesses are those that practice and experience will alleviate. If the Science Court is given adequate funding and can satisfy the requirements of impartiality no matter what its decisions turn out to be, it will have succeeded in bringing order to at least one component of scientific debate. In doing so, it will stand, with the formulation of accepted moral policies, as the cornerstone for a concept of constructive societal interference in scientific affairs, an intervention based not upon fear and hysteria but on reasonable doubts anchored in knowledge.

Other Mechanisms for Gathering Information

Another possible solution to the search for order within scientific debate is the appointment of the Certified Public Scientist (CPS), who, as a person on the public payroll uninvolved in research, would act as a neutral overseer in any disputes arising out of scientific research. Several problems with such an arrangement are immediately evident. A single individual whose full-time profession is to oversee regulation of research is far more susceptible to what might euphemistically be called "influence" than would be a panel of professionals, chosen for a specific task. We already know that federal workers are near the bottom of the pay scale at almost every professional level; in the regulation of an industry with the potential for profit that recombination has, keeping the guardians straight would require more guardians.

Another problem, equally as vexing, might arise when the CPS tries to formulate decisions. Because of the nature of the scientific hierarchy and its emphasis on the godlike qualities of those who have successfully completed dramatic and revolutionary research, it would be almost impossible for such regulators to challenge those taming the frontiers of science.

One solution that would not require the creation of an entirely new, independent branch of government might be to set up a series of commissions, similar to the National Commission for the Protection of Human Subjects, to serve as an alternative to the Science Court. Such commissions would function under the aegis of the government official responsible for the general regulation of research—the Secretary of Health, Education and Welfare, for example, or, for nuclear research, the Secretary of Energy. Unfortunately, as it now stands, the Commission for the Protection of Human Subjects has little power of its own; its sole responsibility is to report to the Secretary of HEW. But the Secretary is legally required to pay at least some attention to the commission's recommendations; if he turns them down, he has sixty days to state his reasons.

Since the main criticisms of this commission are that it has no teeth whatsoever and that its independence is compromised by its total reliance on the Secretary of HEW for funding and for action on its recommendations, perhaps these shortcomings should be addressed. If future commissions are given the power to recommend legislation as well as executive action, some of the necessary independence would be restored; commissions that were ignored by one branch of government would have an alternate outlet for their recommendations beyond their direct superiors.

Whether funds are appropriated for a Science Court, a series of investigative commissions, or a Certified Public Scientist, some action will have to be taken to fill the current void in our decision-making process. While we must avoid a situation where "those who can, do, those who can't, regulate," we must also end the confusion that has shadowed the recombinant DNA debate from the very beginning.

Informing the Public

The age of the computer has placed us on the edge of a fundamental change in voting patterns within the United States. Soon, national referenda on an issue-by-issue basis will be possible; the entire population will be able to gather in front of its television sets, be fed the appropriate data, and respond immediately, through computerized, pushbutton polling, to the question at hand.

Such a situation will create a basic new dilemma even as it solves the problems of how to ensure true participation in the political process. In the past, even the minor inconvenience of registering and going to the polls has effectively limited actual participation to those who were willing to sacrifice the time and energy to register to vote and to appear at designated hours. As tiny a limitation as that might seem, it effectively restricted the actual process of election to 30–60 percent of the electorate. Those who did vote were more likely to have been informed about the issues at stake; because the act of voting required some voluntary action on the part of the voter, it was likely that those who voted had also taken some time to become informed, however poorly, about issues and candidates.

Now, all of a sudden, everyone who owns a television set will be enfranchised. And while the logistical problems of getting to the voting booth have, in the past, been little enough insurance that each participant understood the issues at stake, opening the process to the millions who purchased their sets for the sole purpose of watching *Monday Night Football* or *American Bandstand* will remove even that small check from the process. With involvement reduced to the simple push of a button, and with the electorate able to voice its opinions directly on an issue-by-issue basis, rather than merely through its elected representatives, it becomes crucial to discover another way to ensure that the public has the opportunity to be informed, not in the haphazard ways possible today, but in ways commensurate with the powers of computerized communications.

There is, of course, little chance that the future will see legislative bodies replaced by an involved voting public. Officials are elected not only for their legislative and executive expertise but because they take the everyday business of governing off our backs. But the larger problems confronting our political system—the fundamental dilemmas posed by a certain radical line of scientific research, perhaps, or by the desirability of a war, a treaty, or higher taxes—may all soon be placed in front of the public. When this happens the public's duty to learn about the issues will be paramount. The flow of information from the scientific community to the public is, moreover, a special problem. For scientific news is not like ordinary news; it must be filtered through more specialized channels to become accessible to the public. We must therefore ask ourselves two questions: What alternatives exist to the corporate control of news that now exists, to the scoop-for-profit-and-survival motives of most reporters, to the hunger for form rather than substance that has turned most organs of information, both on the air and in print, into carbon-copy redundancies? Would an independent, government-supported information agency, constructed along the lines of the Science Court, be an objective source of information, competitive with the existing news organs?

There is something inherently dangerous in the present motif of news-for-profit. At the very least, it means that the news media are highly susceptible to manipulation from almost any source, that arguments are won or lost not on their merits but on the skill of the public-relations people supporting the adversaries. At its worst, it generates an implied covenant between newsmakers and reporters, a scratch-for-scratch relationship, whereby news is leaked to those who have kept the faith in the past. It is easy to accept that kind of relationship with a wink and a shrug, to chalk success up to someone's cleverness and hold out for the inherent strengths of pragmatic politics. It is slightly more difficult to recognize that the system of dissemination of information as it now stands, while producing dramatic and telling stories in such photogenic sectors of society as politics and the frontiers of research, shortchanges equally important, but less attractive, areas. Recombination, for example, has surfaced in the media as a pitched battle between those who want the power to create another Andromeda strain and those who do not. Published reports on hearings and debates that have been almost exclusively devoted to the questions of ethics and to the future of science in our society have abandoned the central themes to dally in the fascination of scenarios of horror. While one perspective might view this unequal distribution of news space as perfectly fine (after all, the information is available; anyone who really wants it can find it buried in the stacks of almost any library), another might question the wisdom of letting an important issue turn on a publisher's or news director's devotion to the bottom line.

The challenge, then, is both urgent and unchanged; we must find ways to bridge the chasm separating the scientist from the layman. We must develop new and effective techniques for translating the language of science into the dialect of the street.

As the interval between discovery and technology shrinks, the need for an adequate, informed public response increases. The line between success and failure, between our control of technology and its mastery of us, becomes finer by the day. Unfortunately, there is little we can do to compel people to understand the workings of science for their own protection, for ignorance has become but another in the expanding lexicon of permissible states and attitudes for consenting adults. The key to strengthening the chain that links science to society, however, is information. And while there is no way that society can be forced to use what is available, there are media through which scientific information can be as accessible as the next featured rape-murder, kidnapping, or exposé of political bed-hopping. In our society, the separation of media and state has been inviolate; the First Amendment guarantees the press the right to print—and keep out of print—whatever it pleases. But it does not preclude the government from setting up its own mechanism of communication, in the form of a completely independent information agency designed to do nothing more than act as a permanent arm of a Science Court and having the ability to publicize in a commercial way the facts behind scientific disputes.

Genetics is but the latest in a long line of scientific assaults against the combined weight of nature, deprivation, and society itself. With each skirmish of the past, with each radical transformation of our environment, we have had to learn to adapt and to become comfortable with the accelerated progress that no creature born of evolution—except perhaps the cockroach—has ever before been able to withstand. It is a measure of our partial success that we are still alive. But now the threats to our existence have grown stronger. The old battle, between man and nature, has been replaced by a new one, between man and the products of his own mind. It is us against us, and nature is nothing more than the patch of territory over which we are fighting.

The switch in adversaries has forced us to alter our tactics. Nature was, in a sense, easy prey. It was too powerful to be destroyed, too consistent to fool us for long. Our victories gave us new power and knowledge; our losses were little more than temporary setbacks. And so we could attack it with abandon, throwing everything we had into the fray, knowing we had little to lose and much to gain.

Now we are faced with our own creations, the mirror images of ourselves, as devious and dangerous and unpredictable as we are. All of a sudden, we have to be careful. We have something at stake. Nature, the prize, can be torn apart if the game gets too rough. And as nature goes, goes our survival.

What is required is not a new set of uniforms or a cosmetic facelift, but an entirely new plan, one that takes into consideration all of the new variables, one that protects us as it permits us to continue our progress. There is no longer a question of our intervening in nature; our interventions in the past have been too blatant. There is only the question of controlling our own compulsion for excess until we can be certain of the results. We have taken control of our environment and can play the role of gods; we must now recognize that even gods must be careful. The swords they wield also have two edges. And gods only continue to exist as long as they have a place in the minds of men.

Postscript

At 11:47 P.M. on July 25, 1978—just as this book was going to press—Louise Brown, the first test-tube baby, was born. The announcement of her imminent arrival touched off the fervor usually reserved for moon landings or world wars; newspaper and magazine reporters, television and radio crews, and hoardes of curious bystanders practically laid siege to the tiny English hospital in which she was to be born, trying to bribe its personnel and jockeying for each new sliver of information. The Browns' own story was eagerly pursued; finally, rights to it were sold to London's *Daily Mail*, which paid $565,000 for the privilege of disclosing whether the pregnant mother had had eggs or oatmeal for breakfast.

Despite the commotion, the scientific breakthroughs that led to Louise Brown's conception and birth, while dramatic and powerful, were hardly miraculous. In the best scientific tradition, Dr. Patrick Steptoe and Dr. Robert Edwards, the pioneering scientists, merely took existing knowledge and nudged it upward a crucial notch.

The methods used by the two scientists, called "embryo transfer," were actually quite simple. Steptoe, a skillful gynecologist, removed a mature ovum from one of Mrs. Leslie Brown's ovaries. Then, using a culture devised by Edwards that imitated the conditions of the female uterus, the two scientists mixed sperm and egg. The process worked; the egg became fertilized. Two days later, Steptoe implanted the embryo in Mrs. Leslie Brown's womb. (Contrary to what some believe, Louise Brown is not a clone. Her genes are naturally derived, half donated by her father, half by her mother. Nothing that Steptoe and Edwards did advances any further the scientific knowledge or techniques for the cloning of a human being.)

Louise Brown's conception and birth were indeed triumphs of scientific investigation. But at the moment when Steptoe's and Edwards' success was being splashed across newspapers throughout the world, the results of another test-tube experiment—one that had taken place just under five years earlier in New York City—were being challenged in a courtroom.

Doris Del Zio had been diagnosed as infertile; her Fallopian tubes—which normally transport the ovum from the ovaries to the uterus—were completely blocked. In 1973, Mrs. Del Zio went to New York's Columbia Presbyterian Hospital, where Dr. Landrum Shettles practiced medicine and conducted research. Dr. Shettles was deeply involved in attempts to implant the first test-tube baby. Doris Del Zio, with functioning ovaries and blocked Fallopian tubes, was an ideal candidate for his work.

Mrs. Del Zio underwent the minor surgery necessary to remove a single ovum from one of her ovaries; sperm from her husband, John, fertilized the egg. But two days later, before the embryo could be reimplanted in Mrs. Del Zio, another researcher from the hospital, Dr. Raymond Vande Wiele opened the test-tube and destroyed its contents, without first talking to either Dr. Shettles or the Del Zios. The Del Zios later sued.

In the trial the Del Zios contended that Columbia Presbyterian and Dr. Vande Wiele had deprived them of their property. They also claimed that substantial damage had been done to Mrs. Del Zio's psyche. But the real issues, in the eyes of many, involved the ethics of using science as a tool of creation, of artifically concocting life *in vitro*, of, again, playing God.

The Del Zios eventually won part of their case; the jury decided that they had not been deprived of their property, but that Mrs. Del Zio had indeed been psychologically injured by Dr. Vande Wiele's actions, and they awarded her $50,000. But the ethical issues remained unresolved.

The juxtaposition of the Browns' success and the Del Zios' failure is both ironic and appropriate. Side by side, they stand as symbols of both the vast potential of biological research and of the risks, of the magic of modern science and of the dangers. In fact, the specter of testtube babies raises the same equation of benefit and risk, tempered by fundamental questions of moral policy, that are raised by recombinant DNA research. The benefits of test-tube technology are clear. First, mothers left infertile by blocked Fallopian tubes can now bear children. Further, the successful techniques of Dr. Steptoe and Edwards leave us with the chance to discover more about the embryonic state of life, about the first stirrings of fertilized ova; they can help us better understand the causes and prevention of birth defects and infertility; and, ironically, they may even make it easier for us to develop more effective methods of contraception.

But life conceived in a test tube can be as risky as it is exhilarating. Tiny embryos are extremely delicate; conditions must be almost perfect for them to develop normally in the first few days of life. In less skillful hands, the new techniques may deliver not healthy babies but genetic "monsters," in whom the natural process of growth and development has been severely disrupted, and whom the possibility of living as a functioning part of society is almost nonexistent. Suddenly, in the midst of the Browns' joy, the same ethical questions we have already raised come again to the fore. If a genetic disaster does occur, who will be responsible: the researcher, the parents, or the government? The successes give us few problems; but who will take care of the failures? And who will decide what kinds of babies shall be born and what kind shall die? Where, again, do we draw the line?

Both the existence of Louise Brown and the destruction of the Del Zio embryo raises other important issues as well. Dr. Landrum Shettles worked privately, but not secretly; as a result, his research was vulnerable to action by those with opposing senses of ethics. But Drs. Steptoe and Edwards performed their work in strict secrecy, a decision they made at least partly because of the public and professional furor that had accompanied attempts at similar manipulations. Because they were successful —and because their goals were demonstrably benign— they have been hailed as heroes; and the few voices complaining about their "almost fanatical insistence on secrecy"[1] have scarcely been heard. But success obscures the implications. Scientists often work privately, away from the prying of public scrutiny. Given the opportunity (and, in their perceptions, the necessity), they will continue to do so. The biological revolution, with all its promise and peril, is upon us. And unless we begin now to prepare for its orderly progress, there is no way for us to control it.

[1] *Time* Magazine, July 31, 1978, p. 59.

Glossary

adenine—One of the purine bases of nucleic acids.

amino acids—The building blocks of proteins. There are twenty common amino acids in all; they are joined one at a time, like beads on a string, as the final step in the translation of DNA's genetic message.

aminocentesis—The extraction of cells and fluid from the amnion, which surrounds the fetus in the uterus. Growing these cells in culture makes it possible to monitor the fetus' health.

bacteriophage (phage)—A virus that multiplies in bacteria. Bacteriophage lambda is commonly used as a vector in recombinant DNA experiments.

chromosomes—Threadlike structures composed of DNA and protein. With the exception of the free-floating plasmids, they contain all the cell's DNA.

clone—A group of genetically identical cells or organisms asexually descended from a common ancestor. All cells in the clone have the same genetic material and are exact copies of the original.

codon—A sequence of three nucleotides; a unit of nucleic acid which holds the code for one amino acid.

cytosine—One of the pyrimidine bases of nucleic acids.

DNA—Deoxyribonucleic acid. The genetic material found in all living organisms. Every inherited characteristic, every genetically produced function, has its origin somewhere in the code of each individual's complement of DNA.

DNA ligase—An enzyme which can repair breaks along a length of DNA. Ligase is used to anneal together separate strands of DNA during recombination.

dominant gene—A sequence of DNA; a characteristic whose expression prevails over alternative characteristics for a given trait.

enzyme—A functional protein which catalyzes a chemical reaction. Enzymes control the rate of metabolic processes in an organism.

E. coli—A bacterium which commonly inhabits the human intestine. *E. coli* is hardy and promiscuous, factors which make it the organism of choice in many microbiological experiments.

eukaryote—A complex organism, composed of cells with nuclei. All higher organisms, including man, are eukaryotes.

gene—The hereditary unit; a segment of DNA. In a limited sense, a gene is the portion of DNA that controls the expression of a single trait.

genome—A single set of chromosomes.

guanine—One of the purine bases in nucleic acids.

host cell—A cell in which a virus grows and reproduces. In recombinant DNA experiments, the host cell is usually a bacterium, like *E. coli,* into which a virus containing hybrid DNA is inserted.

nucleotide—The fundamental unit of nucleic acids. Nucleotides are classified as acids. They consist of one of the four bases—adenine, guanine, cytosine, and thymine (uracil in the case of RNA)—and its attached sugar and phosphate.

oncogenic—Cancer-causing; tumor-inducing.

one-gene/one-enzyme hypothesis—The principle which states that a single gene directs the synthesis of a single enzyme.

plasmid—Any hereditary material that is not a part of a chromosome. Plasmids are circular and self-replicating. Because they are generally small and relatively simple, they are used in recombinant DNA experiments as acceptors of foreign DNA.

prokaryote—A simple, single-celled organism which contains no nucleus. Its biochemical reactions also differ from those of eukaryotes. All bacteria are prokaryotes.

protein—A large molecule composed of amino acids. Proteins, in structural and functional forms, are both the building blocks and the catalysts for change in living organisms.

purine—A class of organic bases found in nucleic acids. Adenine and thymine (uracil in RNA) are purines.

pyrimidines—A class of organic bases found in nucleic acids.

Guanine and cytosine are pyrimidines.

recessive gene—Any gene whose expression is dependent upon the absence of a dominant gene.

recombination—The independent reassortment of genes, leading to the development of a new organism containing genes other than those of its parents. Recombination has always been a natural occurrence in nature; now it can be controlled by man.

recombinant DNA—The hybrid DNA produced by the process of recombination.

restriction enzyme—An enzyme within a bacterium that recognizes and degrades DNA from foreign organisms, thereby preserving the genetic integrity of the bacterium. In recombinant DNA experiments, restriction enzymes are used as tiny biological scissors to cut up foreign DNA before it is recombined with a vector.

RNA—Ribonucleic acid. In its three forms—messenger RNA, transfer RNA, and ribosomal RNA—it assists in translating the genetic message of DNA into the finished protein.

ribosomes—The parts of a cell which are the sites of the construction of proteins.

tetranucleotide hypothesis—An early theory about the structure of DNA, proposed by P.A. Levene. The tetranucleotide hypothesis stated that the four bases of DNA were present in equal proportions in the DNA molecule. This hypothesis had a profound effect upon early theories about DNA's function, since, if it were true, DNA could hardly be complex enough to contain genetic messages.

thymine—One of the pyrimidine bases of DNA.

transduction—The process by which foreign DNA becomes incorporated into the genetic complement of the host cell.

uracil—One of the pyrimidine bases of RNA.

vector—An organism that transfers a substance from one host to another. In recombinant DNA experiments, the most common vectors are plasmids and phages.

virus—An infectious agent which requires a host cell in order for it to replicate. It is composed of either RNA or DNA wrapped in a protein coat.

Selected Bibliography

Books

Biomedical Research and the Public. Prepared for the Subcommittee on Health and Scientific Research of the Committee on Human Resources, United States Senate, U.S. Government Printing Office, Washington, D.C., 1977.

Cooke, E. Mary. *Escherichia Coli and Man.* Edinburgh and London: Churchill Livingstone, 1974.

Dobzhansky, Theodosius. *Genetic Diversity and Human Equality.* New York: Basic Books, 1973.

Dubos, Rene J. *Science and Utopias.* New York: Columbia University Press, 1963.

Englehart, H.T., and Callahan, D., eds. *Science, Ethics and Medicine.* Hastings-on-Hudson, N.Y.: The Hastings Center, 1976.

Etzioni, Amitai. *Genetic Fix.* New York: Colophon Books, 1973.

Fuller, John G. *We Almost Lost Detroit.* New York: Ballantine Books, 1976.

George, F.H. *Science and the Crisis in Society.* Wiley-Interscience, 1970.

Goodfield, June. *Playing God: Genetic Engineering and the Manipulation of Life.* New York, Random House, 1977.

Hellman, A., Oxman, M.N., and Pollack, R., eds. *Biohazards in Biological Research*. Cold Spring Harbor, N.Y.: Cold Spring Harbor Laboratory, 1973.

Hilton, Bruce, *et al.*, eds. *Ethical Issues in Human Genetics*. New York and London: Plenum Press, 1973.

King, Robert C. *A Dictionary of Genetics*, 2nd ed. New York, London, and Toronto: Oxford University Press, 1972.

Lappé, M., and Morison, R.S., eds. *Ethical and Scientific Issues Posed by Human Uses of Molecular Genetics*. New York: New York Academy of Sciences, 1976.

Milasky, A., and Annas, G.J., eds. *Genetics and the Law*. New York and London: Plenum Press, 1976.

Nyhan, William L. *The Heredity Factor: Genes, Chromosomes, and You*. New York: Grosset & Dunlap, 1976.

Olby, Robert. *The Path to the Double Helix*. Seattle: University of Washington Press, 1974.

Potter, Van Rensselaer. *Bioethics: Bridge to the Future*. Englewood Cliffs, N.J.: Prentice-Hall, 1971.

Rose, Hilary, and Rose, Stephen. *Science and Society*. Baltimore: Penguin Books, 1970.

Sayre, Anne. *Rosalind Franklin and DNA*. New York: W.W. Norton, 1975.

Schooler, Dean. *Science, Scientists, and Public Policy*. New York: New York Free Press, 1971.

Science and Its Public: The Changing Relationship. In *Daedalus,* Summer 1974.

Shannon, J., ed. *Science and the Evolution of Public Policy*. New York: Rockefeller University Press, 1973.

Taylor, G. Rattray. *The Biological Time Bomb*. New York: New American Library, 1968.

Watson, James D. *The Double Helix*. New York: New American Library, 1968.

————. *Molecular Biology of the Gene*, 2nd ed. Menlo Park, Calif.:W.A. Benjamin, 1970.

Articles

Bennett, W., and Gurin, J. "Science that Frightens Scientists," *Atlantic Monthly,* February 1977.

Berg, P., and Singer, M. "Seeking Wisdom in Recombinant DNA Research," *Federation Proceedings,* Vol. 35, no. 14 (December 1976).

Cavalieri, L.F. "New Strains of Life—or Death," *The New York Times Magazine,* August 22, 1976.

Chargaff, Edwin. "On the Dangers of Genetic Meddling," *Science Magazine,* Vol. 192.

Chemical & Engineering News. "News Forum: Recombinant DNA Research: A Debate of the Benefits and Risks," May 30, 1977.

Cohen, S. "The Manipulation of Genes," *Scientific American,* Vol. 233, no. 1 (July 1975).

Curtiss, R. "Genetic Manipulations of Microorganisms: Potential Benefits and Biohazards," *Annual Review of Microbiology,* 1976.

Davis, B., and Sinsheimer, R. "The Hazards of Recombinant DNA," *Trends in Biochemical Science,* August 1976.

Grobstein, C. "The Recombinant DNA Debate," *Scientific American,* Vol. 237, no. 1 (July 1977).

Harvard Magazine, Vol. 79, no. 2 (October 1976).

Judson, H.F. "Fearful of Science," *Harper's,* Vol. 250, no. 1498 (March 1975); "Fearful of Science, Part II," *Harper's,* Vol. 250, no. 1500 (June 1975).

Lubow, A. "Playing God with DNA," *New Times,* January 1977.

Rifkin, J. "DNA," *Mother Jones,* Vol. 2, no. 2, (February/March 1977).

Roblin, R. "Ethical and Social Aspects of Experimental Gene Manipulation," *Federation Proceedings,* Vol. 34, no. 6 (May 1975).

Rogers, Michael. "The Pandora's Box Congress," *Rolling Stone,* June 19, 1975.

Saturday Review, "God and Science: New Allies in the Search for Values," December 10, 1977.

Sinsheimer, R. "Troubled Dawn for Genetic Engineering," *New Scientist,* October 16, 1975.

Wade, Nicholas. Approximately a dozen articles in *Science,* 1975–78.

——. "Go-Ahead for Recombinant DNA," *New Scientist,* Dec. 25, 1975.

Wright, S. "Doubts over Genetic Engineering Controls," *New Scientist,* December 2, 1976.

Documents

Ashby, E., *et al.* "Report of the Working Party on the Experimental Manipulation of the Genetic Composition of Microorganisms." Presented to Parliament, January 1975.

Berg, P., *et al.* "Potential Biohazards of Recombinant DNA Molecules." Letter to the Editor, *Science,* July 26, 1974.

"Guidelines for the Use of Recombinant DNA Molecule Technology, in the City of Cambridge." By the Cambridge Experimentation Review Board, Cambridge, Mass., submitted January 5, 1977.

"NIH Guidelines for Research Involving Recombinant DNA Molecules." U.S. Department of Health, Education and Welfare, June 23, 1976.

"Recombinant DNA Research, Vol. I: Documents Relating to 'NIH Guidelines for Research Involving Recombinant DNA Molecules.'" U.S. Department of Health, Education and Welfare, August 1976.

Singer, M., and Söll. D. "Guidelines for DNA Hybrid Molecules." Letter to the Editor, *Science*, September 21, 1973.

"Summary Statement of the Asilomar Conference on Recombinant DNA Molecules." *Proceedings of the National Academy of Science, U.S.A.*, Vol. 72, no. 6 (June 1975).

Weiner, C., *et al.* "Project on the Development of Recombinant DNA Research Guidelines." Developed by the Massachusetts Institute of Technology, Oral History Program, Cambridge, Mass.

Index